IoT and Wireless Communication Networks

This book series aims to provide a comprehensive overview of the latest developments and best practices in the internet of things (IoT) and wireless communication networking. The series will cover a wide range of topics, including but not limited to AI/ML-based physical layer communication of orthogonal frequency division multiplexing (OFDM), multiple input multiple outputs (MIMO), orthogonal time-frequency space (OTFS), reconfigurable intelligent surfaces (RIS), full duplex, backscatter communication, estimation and detection, signal classification, data pre-processing, feature selection, predictive modeling, anomaly detection, optimization techniques, resource allocation, auto-encoder, and more towards intelligent software defined radio (SDR) and intelligent software-defined network (SDN). The series will be a valuable resource for researchers and practitioners who want to learn about the latest techniques and tools for applying AI/ML to IoT and wireless communication systems. The series will also be valuable for those involved in forming and implementing policies and standards around IoT and wireless technologies. The book titles in this series will provide practical guidance on applying AI/ML techniques to improve the efficiency and effectiveness of these systems, as well as insights into the challenges and opportunities of implementing AI/ML in this context.

G. P. Ramesh · Luca Di Nunzio · N. V. Babu
Editors

Empowering IoT with Distributed Edge Computing

Springer

Editors
G. P. Ramesh
Department of Electronics
and Communication Engineering
St. Peter's Institute of Higher Education
and Research
Chennai, Tamil Nadu, India

Luca Di Nunzio
Department of Electronic Engineering
University of Rome Tor Vergata
Roma, Italy

N. V. Babu
Department of Electronics
and Communication Engineering
SJB Institute of Technology
Bengaluru, Karnataka, India

IoT and Wireless Communication Networks
ISBN 978-981-96-9454-9 ISBN 978-981-96-9455-6 (eBook)
https://doi.org/10.1007/978-981-96-9455-6

This Springer imprint is published by the registered company Springer Nature Singapore Pte Ltd.
The registered company address is: 152 Beach Road, #21-01/04 Gateway East, Singapore 189721, Singapore

About This Book

The rise of the Internet of Things (IoT) has revolutionized the digital landscape, placing distributed edge computing at the forefront of technological advancement. This book, *Empowering IoT with Distributed Edge Computing*, explores the integration of cutting-edge techniques to tackle challenges in IoT environments, including secure data handling, efficient resource allocation, energy optimization, and privacy preservation. Featuring innovative methodologies such as reinforcement learning, advanced optimization algorithms, and neural networks, each chapter addresses practical and theoretical dimensions of IoT-edge solutions. From smart cities to wireless sensor networks, the topics covered offer actionable insights for researchers and industry professionals alike. This book aims to spark collaboration and innovation, providing a resource that supports the development of intelligent, secure, and scalable IoT systems.

Contents

About the Editors

G. P. Ramesh obtained his Ph.D. in electronics and communication from Anna University, Chennai. He also holds an M.E. in applied electronics from Coimbatore Institute of Technology and a B.Tech. from S. V. University, Tirupati. Currently, he works as Professor and Head of the Department of Electronics and Communication Engineering at St. Peter's Institute of Higher Education and Research. Dr. Ramesh has 21 years of teaching experience and has been actively involved in establishing laboratories in the electrical, electronics, and communication engineering departments. He has published over 98 research papers in national and international journals and 92 research papers in national and international conferences. Furthermore, he has supervised 14 Ph.D. students in the field of electronics and communication engineering and is currently guiding eight scholars. Dr. Ramesh has also published three patents and developed three funded products sponsored by MSME.

Luca Di Nunzio received his master's degree (summa cum laude) in electronics engineering and his Ph.D. in systems and technologies for space applications from the University of Rome "Tor Vergata," Italy, in 2006 and 2010, respectively. He has published over 70 research papers in indexed journals. Luca has served on the editorial boards of prestigious research journals and is involved in shaping the scientific discourse and advancing knowledge in his field. Additionally, he serves as a reviewer for renowned scientific journals. He has also been invited as a keynote speaker and track chair for several IEEE conferences. Luca Di Nunzio has held roles as coordinator and principal investigator (PI) in various research projects.

N. V. Babu is currently a Professor in the Department of Electronics and Communication Engineering and the Dean (Academics) at SJB Institute of Technology. With nearly 21+ years of academic experience, he holds a Ph.D. from Kuvempu University, along with B.E. and M.Tech. degrees. Dr. Babu is a Senior Member of IEEE and has published research articles in reputed journals, conferences, and book chapters. He has also served as a reviewer for several indexed journals and holds two granted patents to his credit. He and his team are presently executing a government-funded project supported by the Vision Group on Science & Technology (VGST). Additionally, he has successfully chaired and convened two international conferences in association with IEEE and Springer.

Chapter 1
Tent Chaotic Mapping-Based Zebra Optimization for Feature Selection in Secure Medical Data with Mobile Edge Computing Based on IoT

P. Rachana, S. Rajini, Khalid Nazim Abdul Sattar, Kumar Neeraj, and Alok Kumar Pani

Abstract Mobile edge computing and Internet of Things (IoT) become most general in both private as well as public sectors, playing a progressively crucial role in medical applications. However, they focus on the delay caused by encryption and decryption in transmitting sensitive medical data, especially in systems that need to process large volumes of data quickly. Hence, this research proposes the Tent Chaotic Mapping-Based Zebra Optimization Algorithm (TCM-ZOA) approach for the feature selection process in the collected healthcare medical data. This helps reduce the amount of data to be transmitted by eliminating irrelevant or redundant features. This directly addresses the issue of time consumption in encryption and decryption by reducing data size. The min–max normalization technique is performed in collected healthcare data to normalize the input data. The cryptographic approach of Blowfish is utilized to performing encryption as well as decryption procedure, which provides a greater degree of security against attacks. The experimental results demonstrate that the proposed TCM-ZOA approach attains the better accuracy of 98.39% and encryption

P. Rachana (✉)
Department of Computer Science and Engineering, New Horizon College of Engineering, Bengaluru, India
e-mail: dr.rachanap@newhorizonindia.edu

S. Rajini
Department of Information Science and Engineering, Vidyavardhaka College of Engineering, Mysuru, India

K. N. A. Sattar
Department of CSI, College of Science, Majmaah University, Al Majmaah, Saudi Arabia
e-mail: k.sattar@mu.edu.sa

K. Neeraj
Department of Electronics and Communication Engineering, Anurag University, Hyderabad, India
e-mail: kumarneerajece@anurag.edu.in

A. K. Pani
Department of Computer Science and Engineering, Birla School of Applied Sciences, Birla Global University, Bhubaneswar, India

G. P. Ramesh et al. (eds.), *Empowering IoT with Distributed Edge Computing*, IoT and Wireless Communication Networks, https://doi.org/10.1007/978-981-96-9455-6_1

time of 0.203 s respectively as compared to the existing methods like ZOA and two fish.

Keywords Blowfish · Internet of Things · Mobile edge computing · Tent chaotic mapping · Zebra optimization algorithm

Introduction

Digital advancements named artificial intelligence (AI), 5G networks, Internet of Medical Things (IoMT) and big data analytics involve a reformed challenging disease and the medical illness diagnosis, detection well as treatment (Moqurrab et al. 2022). IoMT with 5G-connected devices provides the medical aids with new as well as modern conveniences. The technological advancements stages in Secure Health Systems (SHA) have been from 1.0 to 4.0 advancements form of a healthcare paradigm. The primary highlights of these paradigm are from the experts to Electronic Health Records (HER), cloud advancements and patients centric, respectively (Singh and Chatterjee 2023; Verma et al. 2022). An IoT-based SHA paradigm is a network of consistent smart devices, examines as well as offers dynamic healthcare solutions to the patients (Alnaim and Alwakeel 2023). In a digital advancement period, the utilization of intelligent devices in everyday is suffering marvellous development in all sectors such as agriculture, medical and so on. IoT-based devices utilize their perception advancements to produce big data and after transfer it through fog or cloud computing to destination on which precise decision-making abilities through applying deep learning (DL) approaches (Jeon et al. 2022). Medicine plans like those for reintegration, diabetes managements as well as Ambient Assisted Living (AAL), have advantageous from a combination of IoMT advancement in various applications (Unal et al. 2022). In condition involving patients with physical injuries, the system has been developed to identify the most efficient medication routine. Nevertheless, in such as condition, remote monitoring of the patient through IoT devices has provided a new dimension to handle the patient's welfare (Al Mudawi 2022). The key highlights of this article are as trails:

- A min–max normalization is considered in a preprocessing step to equalize a user information and to correct similar values.
- The TCM-ZOA approach is proposed as the feature selection technique for the diagnosis of the medical disease in mobile edge computing-based IoT. Chaotic maps like Tent map enhance the algorithm's global search, enabling it to escape local optima.
- The Blowfish algorithm is utilized for the encryption and decryption process. This approach utilizes a 16-round Feistel network structure and extensive key-dependent substitution, which provides a high degree of security against attacks.

The residue of the article is decided as trails: Section "Literature Survey" demonstrates a literature survey. Section "Proposed Methodology" establishes the proposed method. Section "Experimental Results" explains results and discussion. Section "Conclusion" provides a conclusion of this article.

Literature Survey

In this research, the earlier methods based on the mobile edge computing with IoT for the medical and healthcare applications are discussed, along with their benefits and the limitations.

Alrazgan (2022) studied the offloading plans for the mobile edge computing (MEC) as well as resource allocation, where the few intensive computational operations were to be estimated through either nearby edge computing or locally. The optimization approaches such as conventional Particle Swarm Optimization (PSO), Dynamic PSO (DPSO) and Ant Colony Optimization (ACO) were reviewed for the resource allocation. The end-to-end network parts were introduced for the healthcare sectors which were supported to minimize the latency and the scalability. However, resource allocation in MEC environments was challenged due to the dynamic nature of user demands and available resources.

Kaushal et al. (2022) developed the Kernel Homomorphism Twofish Encryption Approach (KHTEA) through Exponential Boolean Spider Monkey Optimization (EBSMO) to enhance a privacy of IoT system. The cutting-edge encryption approach was utilized for the protection of the healthcare data. The normalization techniques utilized for preprocessing the data and the logistic regression with principal component analysis (LR-PCA) were utilized as the feature extraction approach to extract important features and eliminate irrelevant features. In feature selection process, the genetic algorithm (GA) approach was utilized to select the features. However, KHTEA combined with EBSMO for IoT security produced the significant computational overhead.

Gupta et al. (2023) developed the smart healthcare system problems, applications and the services through the design the convolutional neural network (CNN)-based prediction approach with a utilization of edge computing and IoT frameworks. The CNN approach was utilized to estimate the health data gathered through the IoT devices. Moreover, a part of edge strategies was to offer the doctor as well as patients through appropriate health-prediction status through the edge servers. However, CNN model was sensitive to input data quality and variability in data collected through various IoT devices which minimized prediction accuracy.

Gowda et al. (2022) introduced the processing packages to communicate by IoT based on Accident Reduction Model (ARM) as well as Augmented Data Recognition (ADR) model. The authors were designed the attempt to obtain a Quality of Service (QoS) in healthcare by fog computing as well as IoT, focussed the enhancement of various different constraints. The combined hardware system through the software models was introduced for electronic health system's safety through the microcontroller. However, as the number of IoT devices and data points increased, maintaining

QoS was challenged due to enhanced network congestion and higher demand on Fog computing resources.

Humayun et al. (2024) designed a Smart as well as Secure Electronic Health Framework using Cutting-edge Technologies (SSEHCET) which leveraged the potentials of advanced cutting-edge advancements. The machine learning (ML) methods related to the clustering were utilized to perform and determine the irregularity modifications in the patient's data. A blockchain advancement makes sure a secrecy of the complex data of the members. The 5G development was made to enhance the effectiveness of the data transmission and the hand-crafted blockchain with the lightweight Blockchain Consensus Mechanism (BCM) was utilized to ignore an expose of complex healthcare data. However, the reliance on 5G for efficient data transmission was constrained the framework's applicability in areas with minimum 5G infrastructure.

Proposed Methodology

In this research, the various important technologies are considered such as IoT, Mobile edge computing and TCM-ZOA approach for the accurate identification and prediction of the healthcare data. The mobile edge devices are associated to IoT sensor nodes for the transmission of the data in rapid way among the doctors as well as patients. A part of edge server is to deliver a quick reply as well as greater bandwidth utilization through the data transmission. Figure 1.1 establishes a system of the TCM-ZOA method.

System Model

This research proposes the network architecture which associates the cloud data centre with mobile edge computing to attain the better accuracy. Assuming that the mobile edge computing is embedded into the conventional base station. Furthermore,

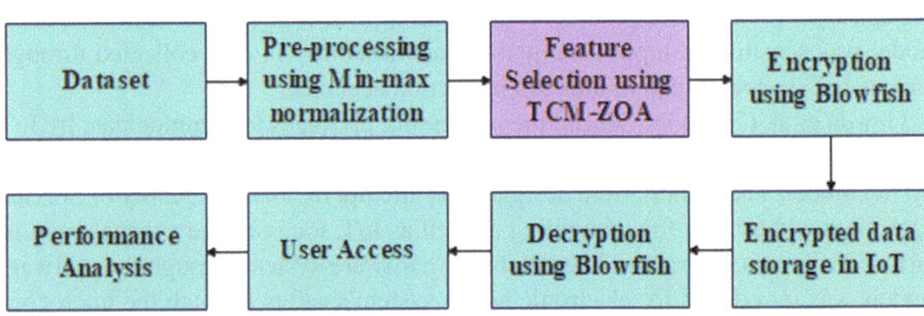

Fig. 1.1 System of the TCM-ZOA method

the base station will be divided into edge computing resources. The function of the edge computing provides the resources required to perform the task offloading. Moreover, the edge computing provides the resource management of an external system, conservation, environmental monitoring as well as operation.

Mobile edge computing is a monitoring of an environment as well as management of the system properties. It is accountable for associating the unique constituents of the network as well as maintaining the taken path data to obtain the target destination. An appropriate approach is utilized to balance the network constituent's loads. Mobile edge computing considers every cloud data centre pool's type as well as quantity of the data necessities when optimizing its cache device. An administration of the cache as well as coordination is important functions of edge computing. In spite of the possibility of associating mobile edge computing, not all edge computing is communicated because of the constraints of the transmission distance.

Dataset

Economic, organizational and medical data are identified in primary files of the AIH and APAC folders. Each file format involves the extended factors which divides the individual healthcare action and is imitated accompanying to the important features. Specifically, average urine result over time Body Mass Index (BMI) measurements are identified in a challenging care as well as data files of the weight loss; however, a mortality represented the data element is till only identified in a hospitalized database object (Kaushal et al. 2022).

Preprocessing

As the device dataset involves specific incorrect as well as incomplete data and the duplication in form through personal potential through both datasets, the preprocessing technique is performed. The min–max normalization technique is performed which has the capability to correct the inaccuracies, incorrect data as well as similar data. The min–max normalization is formulated in Eq. (1.1) as follows:

$$X_{\text{norm}} = \frac{X - X_{\min}}{X_{\max} - X_{\min}}, \tag{1.1}$$

where, X_{norm} demonstrates the normalized value; X denotes the actual data; $X_{\min}$ and $X_{\max}$ demonstrates the minimum and maximum value (Pei et al. 2023).

Feature Selection Using Zebra Optimization Algorithm

In this research, the feature selection is performed for the privacy-preserving for the selection of patient's sensitive data. A ZOA is an optimization approach motivated through a social life behaviour of Zebras in landscape. This approach addresses the challenging mathematical problems through modelling the foraging and defensive responses of the zebra to killers with significant convergence speed as well as search effectiveness. A resolution procedure of ZOA is classified into various phases: population initialization as well as fitness calculation, foraging behaviour as well as defending plan (Yang et al. 2024).

Population Initialization and Fitness Calculation. In ZOA, with respect to acquire the better result to a significant issue, an arbitrary group of zebras is produced as an optimal space for an issue as well as an effective zebra individual positions are identified through continuous estimation. The arbitrarily identified initialized zebra position matrix is formulated in Eq. (1.2) as follows:

$$X_Z = \begin{bmatrix} x_{1,1} & x_{1,2} & \dots & x_{1,d} \\ x_{2,1} & x_{2,2} & \dots & x_{2,d} \\ \vdots & \vdots & \vdots & \vdots \\ x_{N,1} & x_{N,2} & \dots & x_{N,d} \end{bmatrix}. \tag{1.2}$$

In Eq. (1.2), X_Z demonstrates a primarily produced zebra population; $x_{i,j}$ illustrates a position of i th zebra in j th dimension; d as well as N demonstrates a dimension of an optimization issue as well as maximum individuals in Zebra group. A fitness estimation is calculated in Eq. (1.3) as follows:

$$F_Z = \begin{bmatrix} F_1 \\ F_2 \\ \vdots \\ F_N \end{bmatrix} \tag{1.3}$$

In Eq. (1.3), F_Z demonstrates a fitness matrix of zebras; F_i denotes a fitness value of i th zebra.

Foraging Behaviour. In this phase, a zebra's positions are updated by hunting behaviour which directed through a best zebra individual and a position update is formulated in Eqs. (1.4) and (1.5) as follows:

$$x_{i,j}^{n1} = x_{i,j} + a_1 \cdot \left(Z_j^P - a_2 \cdot x_{i,j}\right) \tag{1.4}$$

$$X_i = \begin{cases} X_i & F_i < F_i^{n1} \\ X_i^{n1} & F_i > F_i^{n1} \end{cases}, \tag{1.5}$$

where Z_j^P demonstrates a best zebra lowest fitness known as pioneer zebra; X_i^{n1} demonstrates an unknown position of zebra later updating a position in a foraging behaviour; F_i^{n1} denotes the respective fitness value of a position is lesser, an individual zebra exchanges an actual position through an updated position in an initial phase as well as enter the further phase.

Defending Plan. In this phase, a zebra population updates its position based on a defensive plan. This plan is split through different modes M_1 and M_2 as well as arbitrary number a_3 produced through the range [0, 1] identified which shows the behaviour of zebra; an updated calculation is depicted in Eqs. (1.6) and (1.7) as follows:

$$x_{i,j}^{n2} = \begin{cases} M_1 : x_{i,j} + R \cdot (2 - \text{rand} - 1) \cdot (1 - t/T_{\max}) \cdot x_{i,j} & (a_3 \leq 0.5) \\ M_1 : x_{i,j} + \text{rand} \cdot \left(Z_j^A - a_2 \cdot x_{i,j}\right) & (a_3 > 0.5) \end{cases} \tag{1.6}$$

$$X_i = \begin{cases} X_i F_i < F_i^{n2} \\ X_i^{n2} F_i > F_i^{n2} \end{cases}, \tag{1.7}$$

where R demonstrates the artificially set constant; X_i^{n2} demonstrates an unfamiliar position of i th zebra later a location update of hunting phase; F_i^{n2} denotes its respective fitness value; Z_j^A denotes a value of an arbitrarily chosen zebra individual in j th dimension, and the arbitrary one is known as attacked zebra.

Tent Chaotic Mapping Initialization Population. This approach produces the fully random zebra population in an initialization stage of the population. The conventional initialization approach is resulted in a dissimilar distribution of a primary individuals of the zebra and the greater number of a primary zebra position are distant from an effective solution; a ZOA approach is disposed to fall into the local optimum problem in the following solution procedure. Hence, with respect to produce a zebra population which is most effectively provided in an optimal space, this research produces the TCM to prepare the population as well as updated initialization is formulated in Eqs. (1.8) and (1.9) as follows:

$$x_i = r_i \cdot (u_b - l_b) + l_b \tag{1.8}$$

$$r_i = \begin{cases} \frac{r_{i-1}}{\beta} & r_{i-1} \in [0, \beta] \\ \frac{(1-r_{i-1})}{1-\beta} & r_{i-1} \in (\beta, 1] \end{cases}, \tag{1.9}$$

where x_i demonstrates an individual zebra after an initialization of the tent chaos; r_i demonstrates the produced chaotic sequence; u_b and l_b illustrates upper and lower boundaries of produced zebra locations and β is a flexible chaos parameter.

Encryption and Decryption Using Blowfish Algorithm

Through a Blowfish encryption approach (Adeniyi et al. 2022), 1 16-round Feistel procedure is utilized to encrypt the data. The key-subordinate exchange as well as information-subordinate spare presents in every cycle. Every operation is 32-bit XOR as well as augmentations. The Blowfish approach performs the broad variety of techniques. Before performing any encryption or decryption, the secret keys are to be registered prior. The key clusters are frequently called as P-exhibit is designed based on the 18 32-bit subkeys from P1 to P18, respectively.

An encryption needs an operation which iterates a network for 16 number of times. Every round demonstrates the key as well as data-dependent permutation and key data-independent replacement. For 32-bit words, every operation involves the XORs as well as additions. For every cycle, various indexing array data recovery sets are the individual supplementary process. Decryption is indistinguishable to the encryption through an exception that P1, P2, and P18 are utilized as a kind of switch formation. This approach implements which needed a rapid hustle should unfold the loop as well as make sure that all subkeys are kept in a cache. This approach has been shown to be resistant to various general cryptographic attacks like differential and linear cryptanalysis, designing it a robust choice for various encryption requirements. The encrypted data is then stored in the IoT database, and the decryption process is done once the user accesses the data.

Experimental Results

The significance of the proposed TCM-ZOA method with mobile edge computing-based IoT is implemented on Python 3.8 with the system configurations of intel i5 processor, windows 10 OS and 16 GB RAM. The effectiveness of the proposed TCM-ZOA approach is estimated through utilizing various performance metrices named as precision, accuracy, F1-score and recall. The mathematical expressions of these performance metrices are formulated in Eqs. (1.10)–(1.13) as follows:

$$\text{Accuracy} = \frac{\text{TP} + \text{TN}}{\text{TP} + \text{TN} + \text{FP} + \text{FN}} \tag{1.10}$$

$$\text{Precision} = \frac{\text{TP}}{\text{TP} + \text{FP}} \tag{1.11}$$

$$\text{Recall} = \frac{\text{TP}}{\text{TP} + \text{FN}} \tag{1.12}$$

$$\text{F1-score} = \frac{\text{Precision} \times \text{Sensitivity}}{\text{Precision} + \text{Sensitivity}}, \tag{1.13}$$

where TP represents a true positive; TN represents a true positive; FP represents a false positive and FN represents a false negative.

Performance Analysis

An effectiveness of the proposed TCM-ZOA approach is estimated by using various existing approaches. Table 1.1 demonstrates the performance evaluation of security level results. The existing cryptography approaches such as Digital Signature Algorithm (DSA), Rivest Cipher 4 (RC4), Rivest Shamir Adleman (RSA) and Twofish with the proposed TCM-ZOA with Blowfish approach. Through selecting the most relevant features, TCM-ZOA supports to enhance the accuracy of classification algorithms in medical applications. The proposed TCM-ZOA with Blowfish approach attains the better security level of 98.39% respectively.

Table 1.2 demonstrates the performance analysis of encryption, decryption and execution time. The existing cryptography algorithms such as DSA, RC4, RSA and Twofish are estimated and compared with the Blowfish algorithm. Blowfish is efficient in memory usage, designing it optimal for systems with constrained resources, such as embedded systems and IoT devices. The utilized Blowfish approach attains the encryption time of 0.203 s, decryption time of 0.135 s and execution time of 0.481 s respectively.

Table 1.1 Performance estimation of security level results

Methods	Security level (%)
DSA	86.28
RC4	89.81
RSA	93.91
Twofish	95.28
TCM-ZOA + Blowfish	98.39

Table 1.2 Performance analysis of encryption, decryption and execution time

Methods	Encryption time (s)	Decryption time (s)	Execution time (s)
DSA	0.673	0.642	0.874
RC4	0.573	0.573	0.783
RSA	0.422	0.497	0.643
Twofish	0.382	0.232	0.583
TCM-ZOA + Blowfish	0.203	0.135	0.481

Discussion

At this point, the limitations of the existing works and the advantages of the proposed method are analysed. The limitations of the existing methods are: In Alrazgan (2022), resource allocation in MEC environments was challenged due to the dynamic nature of user demands and available resources. In Kaushal et al. (2022), KHTEA combined with EBSMO for IoT security produced the significant computational overhead. The CNN (Gupta et al. 2023) model was sensitive to input data quality and variability in data collected through various IoT devices which minimized prediction accuracy. In Gowda et al. (2022), as the number of IoT devices and data points increased, maintaining QoS was challenged due to enhanced network congestion and higher demand on Fog computing resources. In Humayun et al. (2024), the reliance on 5G for efficient data transmission was constrained the framework's applicability in areas with minimum 5G infrastructure. To tackle these issues, this research proposes the TCM-ZOA approach for the diagnosis of the medical disease in mobile edge computing-based IoT. TCM-ZOA obtains optimal solutions faster because of the organized yet random initial population designed through chaotic mapping. The enhanced convergence speed is significantly beneficial in mobile edge computing environments, where rapid diagnostic results are important for real-time healthcare monitoring.

Conclusion

Healthcare organization provides wide smart, simple applications and services due to the development in data technology as well as mobile edge computing. Medical systems become a crucial part as a labour-intensive to health conditions, a resource to help commercial processes and discriminator in an expanding health system. Hence, this research proposes the TCM-ZOA approach for performing the feature selection in the medical application with the mobile edge computing-based IoT systems. TCM-ZOA leverages the chaotic mapping properties to enhance convergence speed while preserving accuracy. The Blowfish approach is utilized for performing encryption and decryption to secure the patient's medical data. Blowfish utilizes a 16-round Feistel network structure and extensive key-dependent substitution, which gives a greater degree of security over attacks. The experimental results demonstrate that the proposed TCM-ZOA approach attains the better accuracy of 98.39% and encryption time of 0.203 s respectively as compared to the existing methods like ZOA and twofish. The future work will include the hybrid feature selection as well as encryption and decryption approach to enhance the overall model performance.

References

Adeniyi AE, Misra S, Daniel E, Bokolo A Jr (2022) Computational complexity of modified blowfish cryptographic algorithm on video data. Algorithms 15(10):373

Al Mudawi N (2022) Integration of IoT and fog computing in healthcare based the smart intensive units. IEEE Access 10:59906–59918

Alnaim AK, Alwakeel AM (2023) Machine-learning-based IoT–edge computing healthcare solutions. Electronics 12(4):1027

Alrazgan M (2022) Internet of medical things and edge computing for improving healthcare in smart cities. Math Probl Eng 2022(1):5776954

Gowda D, Sharma A, Rao BK, Shankar R, Sarma P, Chaturvedi A, Hussain N (2022) Industrial quality healthcare services using Internet of Things and fog computing approach. Meas Sens 24:100517

Gupta P, Chouhan AV, Wajeed MA, Tiwari S, Bist AS, Puri SC (2023) Prediction of health monitoring with deep learning using edge computing. Meas Sens 25:100604

Humayun M, Alsirhani A, Alserhani F, Shaheen M, Alwakid G (2024) Transformative synergy: SSEHCET—bridging mobile edge computing and AI for enhanced eHealth security and efficiency. J Cloud Comput 13(1):37

Jeon G, Albertini M, Bellandi V, Chehri A (2022) Intelligent mobile edge computing for IoT big data. Complex Intell Syst 8(5):3595–3601

Kaushal RK, Bhardwaj R, Kumar N, Aljohani AA, Gupta SK, Singh P, Purohit N (2022) Using mobile computing to provide a smart and secure Internet of Things (IoT) framework for medical applications. Wirel Commun Mob Comput 2022(1):8741357

Moqurrab SA, Tariq N, Anjum A, Asheralieva A, Malik SU, Malik H, Pervaiz H, Gill SS (2022) A deep learning-based privacy-preserving model for smart healthcare in Internet of medical things using fog computing. Wirel Pers Commun 126(3):2379–2401

Pei X, Hong Zhao Y, Chen L, Guo Q, Duan Z, Pan Y, Hou H (2023) Robustness of machine learning to color, size change, normalization, and image enhancement on micrograph datasets with large sample differences. Mater Des 232:112086

Singh A, Chatterjee K (2023) Edge computing based secure health monitoring framework for electronic healthcare system. Clust Comput 26(2):1205–1220

Unal D, Bennbaia S, Catak FO (2022) Machine learning for the security of healthcare systems based on Internet of Things and edge computing. In: Cybersecurity and cognitive science. Academic Press, pp 299–320

Verma P, Tiwari R, Hong WC, Upadhyay S, Yeh YH (2022) FETCH: a deep learning-based fog computing and IoT integrated environment for healthcare monitoring and diagnosis. IEEE Access 10:12548–12563

Yang B, Liu Y, Liu Z, Zhu Q, Li D (2024) Classification of rock mass quality in underground rock engineering with incomplete data using XGBoost model and Zebra optimization algorithm. Appl Sci 14(16):7074

Chapter 2
Epsilon Greedy Strategy-Based Q Learning for Data Management in Edge Computing-Based Internet of Things

Pushpa Mohan, K. S. Nandini Prasad, K. Durga Bhavani, R. Mohan Naik, and Alok Kumar Pani

Abstract To encounter a progressive Internet of Things (IoT) system requirements, edge computing forwards dispensation energy as well as loading nearest to an edge network to reduce the power or energy consumption. Edge computing is progressively famous due to these benefits; however, it poses difficulties in effectively handling the resources. However, most existing approaches eliminate the flexibility and effectiveness impacted through the firm architecture of the industrial control system as well as 'end-to-end' edge computing network of IoT. Hence, this research proposes the EdgeAISim architecture named Epsilon Greedy Strategy-based Q Learning (EGS-QL) approach with edge computing is proposed for scheduling the tasks and the resource management. The EGS system balances the exploration as well as exploitation through randomly selecting actions with a small probability while mainly selecting actions with high expected rewards. This make sure the learning process doesn't deteriorate in suboptimal solutions and adapts efficiently to dynamic data

Pushpa Mohan (✉)
Department of Computer Science and Engineering, CMR Institute of Technology, Bengaluru, India
e-mail: pushpa.m@cmrit.ac.in

K. S. Nandini Prasad
Department of Information Science and Engineering, Dayananda Sagar Academy of Technology and Management, Udaypura, Bengaluru, India

K. Durga Bhavani
Department of Computer Science and Engineering, SRKR Engineering College, Bhimavaram, India

R. Mohan Naik
Department of Electronics and Communication Engineering, Shri Dharmasthala Manjunatheshwara Institute of Technology, Ujire, India
e-mail: mohannaik@sdmit.in

A. K. Pani
Department of Computer Science and Engineering, Birla School of Applied Sciences, Birla Global University, Bhubaneswar, India

G. P. Ramesh et al. (eds.), *Empowering IoT with Distributed Edge Computing*, IoT and Wireless Communication Networks, https://doi.org/10.1007/978-981-96-9455-6_2

management tasks. The experimental results demonstrates that the proposed EGS-QL approach attains the better total power consumption of 995 W as compared to the existing method of QL with graph neural network (GNN) approach.

Keywords Edge computing · Epsilon greedy strategy · Internet of Things · Q learning · Resource management · Task scheduling

Introduction

In recent periods, mobile edge computing and Internet of Things (IoT) are observed various conditions to accomplish an advanced system prerequisite. The technology growth focusses on a necessity of on data communication, reducing a throughput as well as network load (Vaiyapuri et al. 2022). The mobile devices and IoT are considered part of future devices, and the technologies are changed to familiarize for the above modifications. Together with IoT, various helpful advancements are developing that are significant development for an entire prospective that is machine learning (ML) will allow the mobile device systems to offer 'intelligence' which have the capability to effectively operate the data (Robles-Enciso and Skarmeta 2023; Le et al. 2022). The devices are digitized to the IoT in various manner, especially in business and healthcare. They have often to transmit a prosperity of important data for an efficient deployment of healthcare systems (Rajavel et al. 2022). The procedure of edge learning requires the upload and download of large-dimension ML parameters and their continuous updated between various edge devices (Tripathy et al. 2022). Comprehending the aim of edge learning with maximum communication effectiveness needs advanced approached of new distributed signal processing as well as wireless approaches which seamlessly integrate communications (Xu et al. 2023; Matilla et al. 2022). Conventional IoT data processing as well as examination system handovers a gathered data to a processing centre for estimation as well as repository. Nevertheless, this architecture is never appropriate for IoT where large number of intelligent sensing devices are implemented in an environment, due to the IoT system produces the greater number of data (Zhang et al. 2022). The primary highlights of this research are as pursues:

- The Epsilon Greedy Strategy-based Q Learning (EGS-QL) approach with Edge computing is proposed for scheduling the tasks. An integration of QL and edge computing allows the dynamic adaptation to changing workloads in a data management.
- A EdgeAISim is proposed in this research which solves the difficulties of the task migration as well as resource management in adaptive edge computing situations. Utilizing progressive approaches as well as implementation, EdgeAISim allows the continuous task scheduling as well as workload balancing over edge servers.

- Decisions are enhanced through QL which minimizes the energy consumption in edge devices through effectively allocating resources like computation and storage in terms of learned policies. This is important for battery-powered or energy-constrained edge devices.

The continuation of the paper is arranged as follows: Section "Literature Survey" presents the literature survey. Section "Methodology" demonstrates a proposed methodology. Section "Discussion" illustrates the results and discussion. Section "Conclusion" provides the conclusion.

Literature Survey

Nandhakumar et al. (2024) introduced the lightweight python-based framework named EdgeAISim for simulation as well as exhibiting of AI approaches for resource management in edge computing surroundings. The progressive AI approaches like Multi-Deep Q-Networks, Armed Bandit through Upper Confidence Bound, Actor-Critic and Deep Q-Networks through graphical neural network (GNN) were considered in EdgeAISim for the optimization of power utilization with respect to maintain the task migration into an edge computing environment. The general characteristics of the EdgeSimPy framework was extended as well as introduced the AI-assisted simulation models. However, EdgeAISim faced challenges in simulating highly distributed environments which involved different edge devices with various resource capabilities.

Debauche et al. (2022) presented the distributed edge network targeted to process as well as depot IoT and multimedia data nearest to a data developer. This architecture contributed the greatest response time with respect to encounter the requirements of advanced systems. A new framework of the short supply circuit was established for data transmission inspired through the short supply chains in agriculture. Eliminating of unimportant mediators among a data producer as well as consumer who had enhanced the effectiveness. However, as the number of IoT devices and multimedia data sources developed, the distributed edge architecture faced challenges in scaling to accommodate enhanced data volumes.

Nisansala et al. (2022) introduced the multi-layer architecture which facilitated service necessities of simultaneous and real-time application implementation as well as platform-sovereign disposition. After that, the new platform and appropriate modules were developed with associated AI processing as well as edge computer paradigms considering problems based on the heterogeneity, security, scalability as well as interoperability of IoT devices. Then, every constituent is made to maintain control signals, microservice arrangement, flows of the data as well as resource arrangement to integrate with an IoT application necessities. However, integrating diverse IoT devices with varying communication protocols and data formats faced significant challenges in maintaining seamless interoperability within the three-layer architecture.

Zhao et al. (2023) presented the software-defined industry control framework for an enhancement of a suppleness as well as security of IIoT edge networks. A presented network decoupled a software as well as hardware of industrial devices through the virtualization as well as industrial modelling advancements. Furthermore, the new edge computing was adopted named discrete computing to AI-driven IIoT to obtain the effective immediate effectiveness as well as resource utilization. A computing approach optimized a networking of AI-driven industrial applications combined through the multi-objective optimization scheduling approach. However, decoupling software and hardware through virtualization enhanced system complexity, needed advanced expertise in virtualization technologies.

Latif et al. (2022) developed the lightweight trust management model which maintained a confidence of the device and managed a service level trust together through the Quality of Service (QoS). An approach estimated an entire trust of devices through the utilization of QoS parameters to estimate the devices trust by applied weights. The trust management systems utilized QoS parameters which enhanced the results which was support for the determination of hateful edge nodes in edge computing networks. However, assigned weights to QoS parameters was subjective or context dependent, potentially resulted to inaccurate or biased trust estimations.

Methodology

This section discusses the significant definitions as well as concepts to understand this work. Edge computing is a framework which locates the resources nearest to a network edge, allowing rapid data processing as well as immediate competences. It provides the computing power as well as services closer to the data sources as well as end users, minimizing delay as well as enhancing the application performance. Figure 2.1 demonstrates the architecture of the EdgeAISim.

EdgeAISim Framework

Figure 2.1 demonstrates a framework of EdgeAIsim, in that, the general characteristics of an EdgeSimPy framework are extended. Then, developing a modern AI-based simulation systems for scheduling the tasks, service relocation, energy management, flow scheduling of the network as well as mobility provision for edge computing surroundings. The undeveloped edge cloud network involves pursuing mechanisms.

Base Station. Base station provides the network linkage to mobile computing devices into their service zone. A whole feature map is partitioned through various cells, in that every sink node considers the attention of every cell. The mobile computing system is located in anyplace within a cell is considered to maintain uniform association to a sink node of that cell.

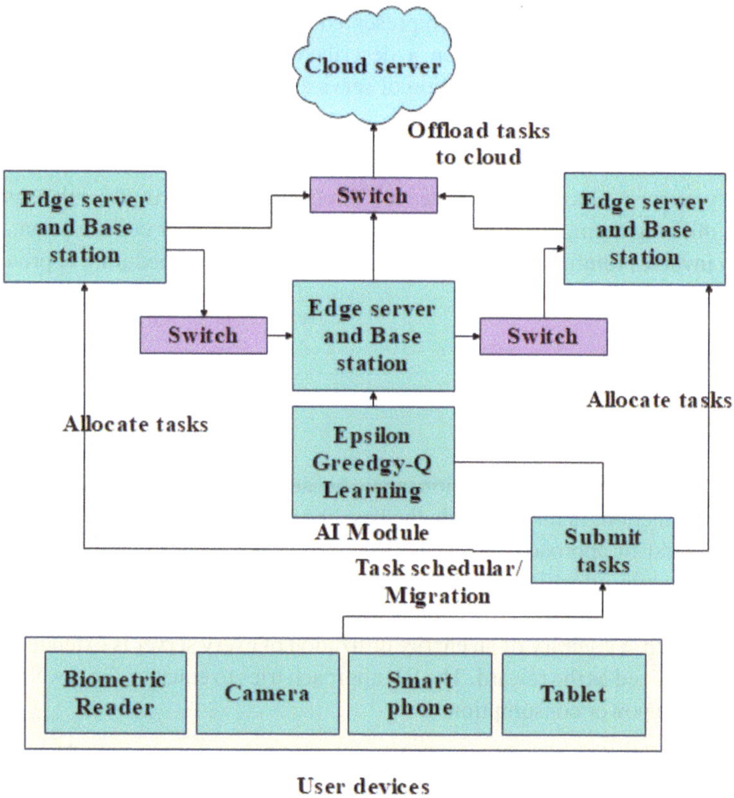

Fig. 2.1 Architecture of EdgeAISim

Network Changes. Network changes are utilized to deliver the influences, generally a wired among the base stations as well as edge servers. The migrations of the tasks are characteristically modelled as network flows, whose timeframe is identified through bandwidth scheduling. A Max–Min normalization approach is utilized for bandwidth scheduling in network changes.

Resource Modelling. Edge servers are utilized to host services. Energy utilization is modelled through different integrated energy consumption approaches: LinearPowerModel, QuadraticPowerModel as well as CubicPowerModel which are based on the various parameters like RAM, CPU as well as hard disk usage. In linear approach, the parameters are related to an extent of 1 polynomial formula. In the quadratic system, the parameters are based on together in an extent of 2 polynomial equation, while in cubic system, the parameters are based upon together in an extent of 3 polynomial equation. Edge servers execute the software systems, allowing them to implement various services at a similar time.

Users. In edge servers, users are represented as the service's consumers who are hosted. Users' forwards based on a described mobility approach, altering their connected sink node for consumption of services placed on an edge server.

Modelling of Tasks and AI Models. Tasks are modelled as the services or the requests which involves certain resource necessities, which are CPU as well as memory request. Once the resources are allocated to the particular edge server, it begins utilizing them, resulting in a modification in its power consumption. An AI system involves reinforcement learning (RL)-assisted task scheduling approach to a significant task allocation.

Task Migration

The task is extended to the migration queue because of the QoS necessities as well as migration function. In this research, the baseline resource management is introduced through worst-fit approach. Then, the EGS-QL is proposed to optimization power utilization while effectively maintain the task migration into an edge computing setting. Every time a migration approach is known as modelled as one timestep. At every timestep, a contrary of an energy utilization of every server is extended as well as total is utilized as the reward. The RL approach tries to enhance the reward as well as reduce the power consumption.

Epsilon Greedy Strategy-Based Q-Learning. QL is a value-based RL approach, which involves various important elements such as environment (E), state (s), action (a) and reward (r) based on a reward function $R(s, a)$ (Zhong et al. 2024; Ma et al. 2022). For particular task, observing a complete state-action-rewards trajectory $T = (s_1, a_1, r_1, \ldots)$, and this process is defined through Markov decision process. Basically, an updated principle of Q-value is estimated in Eq. (2.1) as follows:

$$Q^{\text{new}}(s_1, a_1) = Q\Big((s_t, a_t) + \alpha\Big(r_t + \gamma \max_a Q(s_{t+1}, a) - Q(s_t, a_t)\Big)\Big) \tag{2.1}$$

where α demonstrates a learning rate to manage an augmentation; r_t demonstrates a reward for state s_t after utilizing an action a_t. Deduction factor γ identifies a significance of future rewards and $\max_a Q(s_{t+1}, a)$ denotes maximum reward which attained from state s_{t+1} respectively.

In this research, the online QL approach is adopted to direct a configuration of an optimization sequence and a benefit of an online approach is that the learning procedure as well as optimization of an occurrence occurs concurrently, which saves a computational cost. As per a relationship among the data management and QL, the following concepts are to be involved: an environment in Q-learning respective to fitness scenery issues, agents are the individuals in a data management, a state of an agent denotes a fitness is enhanced or collapsed by an agent, actions are the search strategies which is considered in a collection as well as reward based on the individual is estimated in Eq. (2.2) as follows:

$$R_t = F_{t-1}(S_{t-1}, A_{t-1}) - F_t(S_t, A_t) \tag{2.2}$$

where F_t and F_{t-1} denote the fitness at a generation t and $t-1$ individually. The standard functions in the experiments are reduction issues, once the better solution is acquired, the reward will be positive and contrarywise. For the strategy A, the total reward is formulated in Eq. (2.3) as follows:

$$\overline{R}_t^A = \frac{\sum_{i=1}^{n_A} R_t^A}{n_A} \tag{2.3}$$

where n_A demonstrates the time of the strategy A to be adopted; $\overline{R}_t^A$ denotes the averaging enhancement or reward through the search strategy A. In an action collection, adopting the general ε-greedy plan. Simplify, if the arbitrary value is less than a threshold ε, the data management will select the random operator.

The ε-greedy approach is utilized to choose an action plan. A probability of $1-\varepsilon$ is utilized to choose an action through maximum state action value. A probability ε is utilized to choose an arbitrary action. Eventually, a strategy with greatest cumulative reward value is chosen. An estimation of ε-greedy strategy is formulated in Eq. (2.4) as follows:

$$\text{prob}(a_t) = \begin{cases} 1-\varepsilon, & \text{if } a = \arg\max_{a_t \epsilon A} Q(s, a_t) \\ \varepsilon, & \text{others} \end{cases} \tag{2.4}$$

where $\text{prob}(a(t))$ denotes an agent's selection of action strategy. Table 2.1 demonstrates a hyperparameter value of the proposed method.

Table 2.1 Hyperparameters of the proposed method

Hyperparameter	Value
Learning rate (α)	0.05
Exploration factor ($\in$)	1
Imminent reward weight influence (γ)	0.9
Decay	0.997
Dropout possibility in NN	0.5

Experimental Results and Discussion

An effectiveness of the proposed EGS-QL approach with edge computing-based IoT network is implemented by using Python 3.8 tool with the system configuration of intel i5 processor, 16 GB RAM and windows 10 OS. The performance metrices such as energy consumption.

Performance Analysis

In this section, the significance of the proposed EGS-QL method is estimated with the existing methods related to the different edge servers. The details of the baseline model and the comparative analysis with the discussion are described in the following section.

Baseline Model. Table 2.2 demonstrates a performance evaluation of energy consumption for the six servers through total energy consumption for different algorithms. The existing methods like Actor-critic, RL, QL and greedy strategy-based QL (GS-QL) approach are compared and estimated with the proposed EGS-QL approach. Through leveraging edge computing, the QL algorithm is distributed across edge nodes, minimizing the computational burden on central servers. This is especially helpful for managing large-scale data generated in various IoT systems or distributed environments. The proposed EGS-QL approach attains the better power consumption of 176.3 W, 154.23 W, 178.3 W, 195.2 W, 78 W, 61 on the six different edge servers as well as attains the total power of 995 W, respectively.

Table 2.3 demonstrates a performance evaluation of the latency of the proposed approach based on the different number of tasks. The existing methods like Actor-critic, RL, QL and GS-QL approach are compared and estimated with the proposed EGS-QL approach. The proposed EGS-QL approach attains the minimum latency of 0.75 s, 0.91 s, 1.21 s, 1.52 s and 1.87 s on the different number of tasks like 20, 40, 60, 80 and 100 individually. These results demonstrates that the proposed EGS-QL approach attains the significance results as compared with previous approaches.

Table 2.2 Performance evaluation of energy consumption for different approaches

Methods	Edge server (W)						
	1	2	3	4	5	6	Total power
Actor-Critic	189.2	169.7	193.4	207.3	91	73	1292
RL	185.6	165.4	189.3	204.4	88	71	1045
QL	183.2	161.7	185.0	201.2	85	68	1032
GS-QL	179.4	157.6	182.4	198.7	81	64	998
EGS-QL	176.3	154.23	178.3	195.2	78	61	995

Table 2.3 Performance evaluation of latency (s) based on different number of tasks

Methods	Number of tasks				
	20	40	60	80	100
Actor-Critic	2.01	2.42	2.54	2.65	2.76
RL	1.93	2.12	2.32	2.40	2.53
QL	1.43	1.75	1.91	1.98	2.05
GS-QL	0.91	1.39	1.45	1.67	1.73
EGS-QL	0.75	0.91	1.21	1.52	1.87

Table 2.4 demonstrates the performance evaluation of delay of the proposed method based on the different number of tasks. The existing methods like Actor-critic, RL, QL and GS-QL approach are compared and estimated with the proposed EGS-QL approach. The proposed EGS-QL approach attains the minimum delay of 0.82 s, 0.91 s, 0.95 s, 1.03 s and 1.12 s, respectively. These results demonstrates that the proposed EGS-QL approach attains the significance results as compared to previous approaches.

Comparative Analysis. At this point, the comparative analysis of the proposed EGS-QL with the existing methods are shown. The existing method like EdgeAISim (Nandhakumar et al. 2024) are estimated and compared with the proposed EGS-QL approach. As compared to the existing method, edge computing allows a data processing and decision-making to happen nearer to a data source, minimizing latency. The QL algorithm can respond quickly to changes in data patterns, such as predicting workloads or optimizing storage. Table 2.5 demonstrates the comparative estimation of EGS-QL method with existing method.

Table 2.4 Performance evaluation of the delay (s) based on different number of tasks

Methods	Number of Tasks				
	20	40	60	80	100
Actor-Critic	1.43	1.47	1.51	1.62	1.76
RL	1.22	1.32	1.45	1.53	1.65
QL	1.04	1.09	1.13	1.21	1.34
GS-QL	0.91	0.95	0.98	1.17	1.21
EGS-QL	0.82	0.91	0.95	1.03	1.12

Table 2.5 Comparative evaluation for energy consumption of EGS-QL approach

Methods	Edge server (W)						
	1	2	3	4	5	6	Total power
QL with GNN (Nandhakumar et al. 2024)	181.4	166.2	184.5	200	81	63	1000
Proposed EGS-QL	176.3	154.23	178.3	195.2	78	61	995

Discussion

This section discusses the advantages of the proposed EGS-QL and the drawbacks of the existing works based on an edge computing and IoT network for the data management. The limitations of the existing works are EdgeAISim (Nandhakumar et al. 2024) faced challenges in simulating highly distributed environments which involved different edge devices with various resource capabilities. In Debauche et al. (2022) approach, as the number of IoT devices as well as multimedia data sources developed, the distributed edge architecture faced challenges in scaling to accommodate enhanced data volumes. In Nisansala et al. (2022), integrating diverse IoT devices with varying communication protocols and data formats faced significant challenges in maintaining seamless interoperability within the three-layer architecture. In Zhao et al. (2023), decoupling software and hardware through virtualization enhanced system complexity, needed advanced expertise in virtualization technologies. In Latif et al. (2022), assigned weights to QoS parameters was subjective or context-dependent, potentially resulted to inaccurate or biased trust estimations. Hence this research proposes the EGS-QL approach with the Mobile computing based IoT for the data management system. The epsilon-greedy strategy balances exploration as well as exploitation. This make sure the learning process determines effective data management strategies while exploring diverse possibilities in a dynamic IoT environment.

Conclusion

Edge computing is a developing advancement in a generation, distribution, storage as well as data computation is employed at network edge. The significant apprehension in cloud computing is addressed through edge computing; however, some of the apprehensions like latency as well as energy consumption at an edge and the resources are requires to be solved. This research offerings the flexible edge computing plan for AI-driven industrial IoT. An EGS-QL approach with edge computing is proposed for the effective task scheduling and resource management. Through leveraging edge computing, the QL approach is distributed over edge nodes, minimizing the computational burden on central servers. This is especially helpful for managing large-scale data generated in different IoT systems or the distributed environments. The experimental results demonstrates that the proposed EGS-QL approach attains the better total power consumption of 995W as compared to the existing method of QL with GNN approach. The future work will include the hybrid RL approach for the EdgeAISim with edge computing-based IoT to enhance the overall system performance.

References

Debauche O, Mahmoudi S, Guttadauria A (2022) A new edge computing architecture for IoT and multimedia data management. Information 13(2):89

Latif R, Ahmed MU, Tahir S, Latif S, Iqbal W, Ahmad A (2022) A novel trust management model for edge computing. Complex Intell Syst 8:3747–3763

Le KH, Le-Minh KH, Thai HT (2022) Brainyedge: an ai-enabled framework for iot edge computing. ICT Express 9(2):211–221

Ma T, Lyu J, Yang J, Xi R, Li Y, An J, Li C (2022) CLSQL: improved Q-learning algorithm based on continuous local search policy for mobile robot path planning. Sensors 22(15):5910

Matilla DM, Murciego AL, Jiménez-Bravo DM, Mendes AS, Leithardt VR (2022) Low-cost edge computing devices and novel user interfaces for monitoring pivot irrigation systems based on Internet of Things and LoRaWAN technologies. Biosys Eng 223:14–29

Nandhakumar AR, Baranwal A, Choudhary P, Golec M, Gill SS (2024) Edgeaisim: a toolkit for simulation and modelling of AI models in edge computing environments. Meas Sens 31:100939

Nisansala S, Chandrasiri GL, Prasadika S, Jayasinghe U (2022) Microservice based edge computing architecture for internet of things. In: 2022 2nd International conference on advanced research in computing (ICARC). IEEE, Belihuloya, Sri Lanka, pp 332–337

Rajavel R, Ravichandran SK, Harimoorthy K, Nagappan P, Gobichettipalayam KR (2022) IoT-based smart healthcare video surveillance system using edge computing. J Ambient Intell Humaniz Comput 13(6):3195–3207

Robles-Enciso A, Skarmeta AF (2023) A multi-layer guided reinforcement learning-based tasks offloading in edge computing. Comput Netw 220:109476

Tripathy SS, Imoize AL, Rath M, Tripathy N, Bebortta S, Lee CC, Chen TY, Ojo S, Isabona J, Pani SK (2022) A novel edge-computing-based framework for an intelligent smart healthcare system in smart cities. Sustainability 15(1):735

Vaiyapuri T, Parvathy VS, Manikandan V, Krishnaraj N, Gupta D, Shankar K (2022) A novel hybrid optimization for cluster-based routing protocol in information-centric wireless sensor networks for IoT based mobile edge computing. Wirel Pers Commun 127(1):39–62

Xu W, Yang Z, Ng DWK, Levorato M, Eldar YC, Debbah M (2023) Edge learning for B5G networks with distributed signal processing: semantic communication, edge computing, and wireless sensing. IEEE J Sel Top Signal Process 17(1):9–39

Zhang DG, Ni CH, Zhang J, Zhang T, Yang P, Wang JX, Yan HR (2022) A novel edge computing architecture based on adaptive stratified sampling. Comput Commun 183:121–135

Zhao Y, Hu N, Zhao Y, Zhu Z (2023) A secure and flexible edge computing scheme for AI-driven industrial IoT. Clust Comput 26(1):283–301

Zhong R, Peng F, Yu J, Munetomo M (2024) Q-learning based vegetation evolution for numerical optimization and wireless sensor network coverage optimization. Alex Eng J 87:148–163

Chapter 3
Resource Allocation Using Sine Cosine-Based Egret Swarm Optimization Algorithm in Mobile Edge Computing

N. Sathyanarayana, Supriya, Hirald Dwaraka Praveena, and Mohammed Ziaur Rahman

Abstract Nowadays, mobile edge computing (MEC) is an effective computing model, offers effective computation systems to Internet of Things (IoT). Basically, a placement of MEC servers nearest to the mobile users have significantly minimized the access delays as well as cost of utilizing services. Nevertheless, an edge cloud is mobile and its services are constrained to various neighboring users. Hence, acquiring an ideal offloading policy in the limitations of flexibility and constrained resources poses a complex problem. Thus, this research proposes the Sine Cosine-based Egret Swarm Optimization Algorithm (SC-ESOA) is introduced for an effective resource allocation in MEC. MEC systems are highly dynamic, with different user demands and resource constraints. SC-ESOA's adaptive mechanism which ensures an efficient handling of such variations through modifying resource allocations dynamically. This research also proposes the resource optimization approach for identifying a resource distribution in a system with respect to make sure satisfactory service availability at the less cost. The experimental results show that proposed SC-ESOA attains the minimum energy consumption of 1.65 J and minimum delay of 1.03 s, respectively, as compared to the existing methods like ESOA.

Keywords Edge cloud · Egret swarm optimization algorithm · Mobile edge computing · Resource allocation · Sine Cosine

---

N. Sathyanarayana
Vemana Institute of Technology, Bengaluru, India
e-mail: sathyanarayana@vemanait.edu.in

Supriya (✉)
Department of Information Science and Engineering, Nitte Meenakshi Institute of Technology, Bengaluru, India
e-mail: r.supriya@nmit.ac.in

H. D. Praveena
Department of Electronics and Communication Engineering, School of Engineering, Erstwhile Sree Vidya Nikethan Engineering College, Mohan Babu University, Tirupati, India

M. Z. Rahman
School of Computer Science and Engineering, Presidency University, Bengaluru, India

G. P. Ramesh et al. (eds.), *Empowering IoT with Distributed Edge Computing*, IoT and Wireless Communication Networks, https://doi.org/10.1007/978-981-96-9455-6_3

Introduction

Mobile edge computing (MEC) is a challenging advancement to develop 5G networks as with the development of 5G, low latency high-bandwidth services (Wang et al. 2022a). In an attention of resolving the large complexity, the MEC is developed to prevent the unpredictable burden through aggressing the calculation competences from fundamental to network edges (Zhou et al. 2022). It has arisen as the important framework to manage through an incredible Internet data, where more number of estimation tasks are fulfilled at edge servers in closeness by radio access network (RAN) rather than a detached cloud (Chen et al. 2022a). MEC is the probable idea which places the cloud servers close to the mobile nodes of the mobile networks. Offloading compute processes to the MEC server enhance both quality of computing (Chai et al. 2023). Through effectively regulating the MEC computation as well as bandwidth resources, the task is offloaded from the mobile nods to the closest edge servers to minimize the computation delay as well as local energy consumption, thus promoting an end user's Quality of Experience (QoE) (Jiang et al. 2022). Cloud computing enables the Smart Mobile Device (SMD) to transmit tasks to Edge Cloud Networks (ECN) through high-rapid wireless networks and the Device to Device (D2D) infrastructures with minimum transmission delay as well as battery life (Goudarzi et al. 2023; Tilahun et al. 2022). However, the previous MEC architectures and the divesting devices never fully take into meet an influence of mobility as well as constrained computational resources on offloading strategy (Mahmood et al. 2022). The primary contributions of this research are as follows:

- This research introduces the resource design optimization approach which operated on various network traffic models. This approach targets to enhance a system accessibility while keeping the appropriate QoS fulfilment rate for entire tasks.
- A Sine Cosine-based Egret Swarm Optimization Algorithm (SC-ESOA) is introduced for an effective resource allocation in MEC.
- An integration of sine and cosine functions enhances the balance between exploration and exploitation, enabling the approach to effectively explore the solution space and improve optimal solutions.

Section "Literature Survey" demonstrates the literature survey based on the resource allocation in MEC. Section "Proposed Methodology" outlines the proposed approach. Section "Experimental Results and Discussion" provides the results and discussion. Section "Conclusion" summarizes an overall research.

Literature Survey

In this section, some of the existing works related to the resource allocation based on the mobile edge computing are discussed, along with their advantages and limitations. Amer et al. (2022) introduced the binary phase supportive scheduling approach with the centralized distributed layer. An initial phase was utilized for scheduling the tasks locally on the MEC layers. On the other hand, another level exists at an orchestrator and applied the tasks to the nearest sink or cloud. The resource optimization approach was also introduced to determine the resource distribution in a system with respect to make sure the reasonable service applicability during the less cost. A resource optimization approach involved various dissimilarities which performed based on the traffic model. However, tasks routed through the orchestrator experiences greater latency because of the time required for data transmission and decision-making at the centralized layer.

Liu et al. (2024) presented a multi-objective resource allocation method (MRAM) for IoT. A Pareto Archived Evolution Strategy (PAES) was utilized for identify a resolution group of multi-objective resource allocation plans through minimum execution time, load balance as well as less minimum energy consumption in MEC. Moreover, various criteria decision-making as well as approach for the demand partiality through resemblance to optimal solution were used for acquire an effective multi-objective resource allocation plan. However, the utilization of PAES for multi-objective optimization involved maintaining and explored a large Pareto front, which was computationally expensive.

Yang et al. (2022) initially divided an offloading process into two-step offloading paradigm as well as converse a manner of resolved the offloading decision as well as resource allocation issues to minimize a dimensionality of a state as well as action space. A resource allocation was formulated as the Markov decision process (MDP) as well as utilized the Deterministic Policy Gradient Algorithm (DDPG) to change a load balancing of an edge server. Then, minimized a transmission delay as well as energy and after genetic algorithm (GA) was utilized to identify for the decisions as well as fully connected network (FCN) was utilized to set a decision-making procedure. However, dividing an offloading process into two steps and reversing a sequence created complexity to the framework.

Liu et al. (2023) aimed to reduce an entire implementation cost of various systems through offloading of computations over smart mobile device (SMD) against the edge clouds in Edge Cloud Computing (ECC) structure. Through a consideration of the mobility of the SMD as well as edge clouds, formulated a total cost minimization issue in the limits of the request execution as well as connection time among SMD and edge clouds. The best choice of offloading plan based on the game approach as well as edge cloud payment opposition approach was developed to effectively allocate the edge cloud resources to SMD. However, a framework considered the flexibility of SMDs as well as edge clouds, but rapid movements disrupted the connection and invalidate pre-computed offloading strategies.

Wang et al. (2022b) presented the effective and efficient distributed deep learning-based computation offloading as well as resource allocation (DDL-CORA) approach to Software Defined Mobile Edge Computing (SD-MEC) in IoT. The approach appealed various parallel deep neural networks (DNN) through trained sets of channel improvements to produce judgements of if IoT systems were offloaded a task as well as what small base stations (SBS) to allocate. However, IoT networks were inherently dynamic and the approach does not robustly maintain unpredictable events.

Proposed Methodology

This section describes a proposed framework approach and algorithms which guides a task scheduling issue. SC-ESOA effectively familiarizes to dynamic MEC environments, involving fluctuating workloads and user mobility through adjusting resource allocation. Figure 3.1 demonstrates the proposed architecture.

System Model

This research assumes the ranked system with various layers like end users, base station (BS) or sink prepared through CPU, cloud as well as orchestrator. In Fig. 3.1, an end user layer involves N number of users, all of them is integrated through the sink by large robust signal strength. Because of a constraint of computational systems in an end user layer, tasks are offloaded through users who were integrated with sink. This layer involves *M* sink, every sink schedules the obtained tasks independently on its applicable computation resources. The task which fails to be scheduled through the BS are transmitted to a distributed layer for the global scheduling.

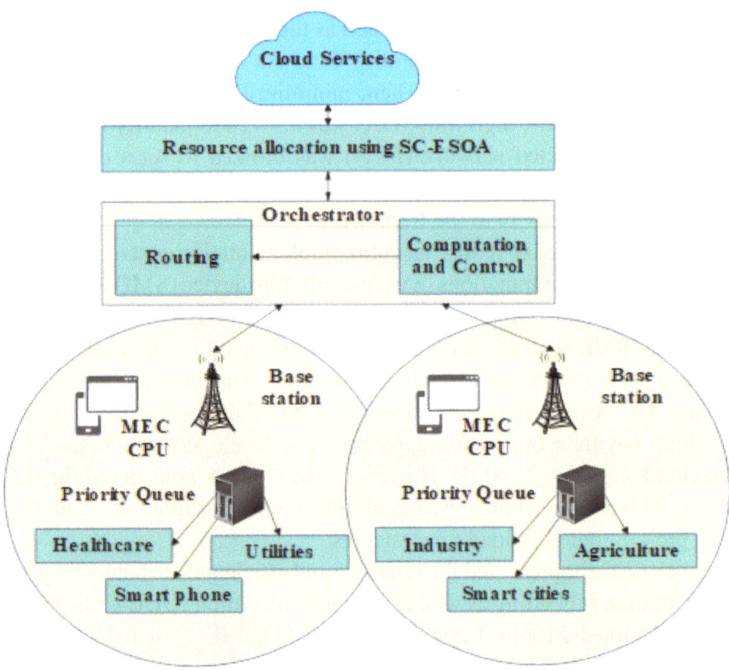

Fig. 3.1 Architecture of the proposed method

A distribution is associated to each sink in that layer in the network area. Also, the distribution system is associated to a cloud layer, that is considered to offer the constrained computation resources which is utilized for the task estimation. The tasks are divided into service kinds which imitate an importance as well as task necessities. An orchestrator schedules the tasks to the nearest sink with permitted computational resources or to a cloud based on task limits, service types as well as network conditions. A system performs in a discrete-time paradigm, which request the existence planned during starting of every time period. A user updated their user-sink integrations in each important cycle, which involves m time periods to model the user mobility. It is considered that every user will transmit an individual service request often. The user is permitted to request an unknown service when it is satisfied or congested. The task is considered as the continuous and to obtain an individual CPU simultaneously, however, it holds a CPU for more than a period.

Resource Allocation

The distribution of CPU over a sink node is an aspect which causes the number of blocked requirements in the network. Nevertheless, eliminating a task blocking rate impacts a user experience to decline and utilizing the maximum number of CPU enhances the expenses as well as energy consumption of the system. Hence, both characteristics are assumed whole designing the distribution of the CPU. In an absence of position approach to acquire the CPU distribution assuming a blocking rate as well as system cost, thus this research proposed to model the CPU distribution as the optimization issue. ESOA approach is proposed to identify the optimal CPU distribution rather than the unreasonable comprehensive search approach.

The number of resources important to offer the acceptable blocking rates are utilized to effectively estimate the scheduling approach. The effective approach uses some resources to contribute an anticipated level of service obtainability.

Egret Swarm Optimization Algorithm

ESOA (Chen et al. 2022b; Alajlan and Razaque 2023) is motivated through a greedy behavior of a snowy egrets as well as large egrets which involves various important strategies such as sit-and-wait, aggressive as well as the discriminant surroundings. Egret population involves n groups of Egret groups and every egret group involves various egrets. Egret A assumes a sit-and-wait plan, Egret B and C individually acquires an arbitrary walk as well as squeeze device in an aggressive plan.

Sit-and-Wait Plan. An opinion Equation of i th Egret A is defined as $\hat{y}_i = E(x_i)$, function E denotes an estimation approach of Egret A for probable target at present location; x_i denotes a location of group i. Estimate a present estimate value of $\hat{y}_i$

through iteration in Eq. (3.1), parameterize a calculation and error e_i is estimated in Eq. (3.2) as follows:

$$\hat{y}_i = w_i * x_i \tag{3.1}$$

$$e_i = \frac{\|\hat{y}_i - y_i\|^2}{2} \tag{3.2}$$

where, w_i denotes the weights of estimation approach; $*$ represents the multiplication; the practical gradient $\hat{g}_i$ of w_i is formulated in Eq. (3.3) and the direction of the egret flight $\hat{d}_i$ is formulated in Eq. (3.4) as follows:

$$\hat{g}_i = \frac{\partial \hat{e}_i}{\partial \hat{w}_i} = \frac{\partial \|\hat{y}_i - y_i\|^2}{\partial w_i} = (\hat{y}_i - y_i) * x_i \tag{3.3}$$

$$\hat{d}_i = \frac{\hat{g}_i}{|\hat{g}_i|} \tag{3.4}$$

where, w_i demonstrates the matrix randomly developed based on the uniform distribution. Egrets will rely on the insights of two egret's predation paths: direction update of best position of egret group (d_h, i), which is formulated in Eq. (3.5) and direction update of the best position in all population $\left(d_g, i\right)$, which is formulated in Eq. (3.6) as follows:

$$d_h, i = \frac{x_{\text{ibest}} - x_i}{|x_{\text{ibest}} - x_i|} * \frac{f_{\text{ibest}} - f_i}{|x_{\text{ibest}} - x_i|} + d_{\text{ibest}} \tag{3.5}$$

$$d_g, i = \frac{x_{\text{gbest}} - x_i}{\left|x_{\text{gbest}} - x_i\right|} * \frac{f_{\text{gbest}} - f_i}{\left|x_{\text{gbest}} - x_i\right|} + d_{\text{gbest}} \tag{3.6}$$

where d_{ibest} and d_{gbest} denote a flight direction of an optimal individual in an egret group and population. x_{ibest} and x_{gbest} denote an optimal value of egret group as well as population; f_{ibest} and f_{gbest} denote a best fitness (energy consumption) of egret group as well as population. Estimate a pseudo gradient of a weight in an opinion formula. An updated Egret A's location is formulated in Eq. (3.7) as follows:

$$x_{a,i} = x_i + \exp\left(\frac{-t}{0.1 * t_{\max}}\right) * 0.1 * \text{hop} * g_i \tag{3.7}$$

where, t demonstrates present number of iterations; $t_{\max}$ demonstrates number of iterations; hop means the D-value of a location boundary, such that a value of hop is similar to an upper constraint of solution of minimum constraint of a solution.

Aggressive Plan. An aggressive strategy acquired through Egret B is random walk. Though an energy utilization of arbitrary walk is large, Egret B is probable to acquire greater advantages. A location update of Egret B is formulated in Eq. (3.8) as follows:

$$x_{b,i} = x_i + \tan(r_{b,i}) * \text{hop}/(1+t) \tag{3.8}$$

where, $r_{b,i}$ denotes an arbitrary number. Egret C acquires a squeeze plan. When it identifies its prey, it will pursue it till it is hunt, which is formulated in Eq. (3.9) as follows:

$$x_{c,i} = (1 - r_i - r_j) * x_i + r_i * D_h + r_j * D_g \tag{3.9}$$

where, D_h demonstrates a variance matrix among a present position as well as an optimal position of egret group; D_g demonstrates a variance matrix among present position as well as effective location of egret position; r_i and r_j denote the random numbers.

Discriminant Condition. Later an every Egret in an egret group estimates an updated location, it will combinedly identify an updated position of an egret group. A matrix solution is formulated in Eq. (3.10) as follows:

$$x_{s,i} = [x_{a,i}, x_{b,i}, x_{c,i}] \tag{3.10}$$

The Egret group compared an updated position as well as fitness of three egrets through which of a prior iteration. If updated position of the one Egret's is superior as compared to a prior iteration, which acquires an update. If an updated position of all egrets is poor as compared to the prior one.

Sine Cosine Algorithm. SC is an arbitrary optimization approach which involves few parameters, greater flexibility, easy principle, implementation as well as simply applied to the optimization issues in various areas. SC makes utilization of accurate assets of sine as well as cosine function. Through effectively modifying a global exploration as well as local deployment capability of an amplitude balance approach of sine cosine function in a search process, an optimization procedure is classified into different phases. In an exploration phase, an optimization approach rapidly identifies an achievable area in a search space through integrating an arbitrary solution. In a development phase, an arbitrary solution will slightly modify as well as change their speed of an arbitrary solution which is lower than that of an exploration phase. A position update equation is formulated in Eq. (3.11) as follows:

$$x_i^{t+1} = \begin{cases} x_i^t + r_m * \sin(r_n) * \left| r_p x_{\text{ibest}}^t - x_i^t \right| r_q < 0.5 \\ x_i^t + r_m * \cos(r_n) * \left| r_p x_{\text{ibest}}^t - x_i^t \right| r_q > 0.5 \end{cases} \tag{3.11}$$

where, x_i^t demonstrates a location of i th dimension of present individual in development t; r_n, r_p, and r_q demonstrate the arbitrary number $[0, 2\pi]$, $[0, 2]$, and $[0, 1]$ respectively; parameter r_m denotes a position field of next solution is inside internal or external a present optimal solution. As the number of iterations enhances, a value of r_m are slightly reduce, which supports to improve a local progress capability of an approach. r_q means an arbitrary as well as stable substituting sine and cosine function of an adaptation likelihood.

Experimental Results and Discussion

In this research, various experiments are performed to estimate an effectiveness of the proposed SC-ESOA approach. Particularly, the simulation experiment is implemented on Python 3.10.11 tool with the system configuration of Intel i5 processor, 16 GB RAM and Windows 10 OS. Table 3.1 demonstrates the system configurations of the SC-ESOA method.

Performance Analysis

The success of the proposed SC-ESOA approach is estimated and compared through different resource allocation approaches. Table 3.2 demonstrates the performance evaluation of energy consumption of the existing methods based on number of IoT applications. The existing methods like Remora Optimization Algorithm (ROA), Grey Wolf Optimization (GWO), Pelican Optimization Algorithm (POA), and ESOA are estimated and compared with the proposed SC-ESOA approach. Through allocating resources significantly, the ESOA reduces the energy consumption and operational costs for both mobile devices and edge servers. The proposed SC-ESOA approach attains the minimum energy consumption of 1.65 J, 2.33 J, 3.57 J, and 4.29 J on the number of IoT applications of 100, 200, 300, and 400, respectively.

Table 3.1 System configurations of the SC-ESOA approach

Parameters	Values
Number of virtual machine on every MEC server	8
Power rate of MEC server	350 W
Power rate of performed VM occurrences	75 W

Table 3.2 Performance evaluation of energy consumption (J) based on number of IoT applications

Methods	Number of IoT applications			
	100	200	300	400
ROA	6.78	6.43	7.47	8.43
GWO	6.33	5.39	6.26	7.79
POA	5.29	4.92	5.38	6.39
ESOA	2.38	3.56	4.20	5.35
SC-ESOA	1.65	2.33	3.57	4.29

Table 3.3 Performance evaluation of delay (s) based on number of IoT applications

Methods	Number of IoT applications			
	100	200	300	400
ROA	4.29	6.39	7.12	8.39
GWO	3.83	5.13	6.12	7.39
POA	2.22	4.29	5.23	6.39
ESOA	1.32	3.32	4.29	5.28
SC-ESOA	1.03	2.19	3.29	4.13

Table 3.3 demonstrates the performance evaluation of the existing methods based on number of IoT applications. The existing approaches like ROA, GWO, POA, and ESOA approach are estimated and compared with the proposed SC-ESOA approach. The different number of IoT applications like 100, 200, 300, and 400 are considered in this research to estimate the effectiveness of the proposed method. The proposed SC-ESOA approach attains the minimum delay of 1.03 s, 2.19 s, 3.29 s, and 4.13 s, respectively.

Discussion

At this point, the advantages of the proposed SC-ESOA method and the limitations of the existing works are discussed. The limitations of the existing works are: In Amer et al. (2022), tasks routed through the orchestrator experiences greater latency because of the time required for data transmission and decision-making at the centralized layer. In Liu et al. (2024), the utilization of PAES for multi-objective optimization involved maintaining and explored a large Pareto front, which was computationally expensive. In Yang et al. (2022), dividing a offloading procedure into two steps and reversing a sequence created complexity to the framework. In Liu et al. (2023), the framework considered the mobility of SMDs and edge clouds, but rapid movements disrupted the connection and invalidate pre-computed offloading strategies. In Wang et al. (2022b), IoT networks were inherently dynamic and the approach does not robustly maintain unpredictable events. Hence, to overcome to issues, this research proposes the SC-ESOA approach for the resource allocation in MEC. The mathematical properties of sine and cosine functions help in avoiding stagnation in local optima and speeding up convergence to an optimal solution, important for real-time resource allocation in MEC environments.

Conclusion

This research proposes the SC-ESOA approach for allocating the resources in MEC, which effectively minimizes the energy consumption and delay. SC-ESOA optimizes the distribution of computational resources across edge servers, make sure minimal delays and optimal utilization of available resources in MEC systems. The tasks are initially allocated on their local sinks till no resource are accessible. Unallocated tasks are transmitted to an orchestrator, applying a task to nearest base station or the cloud. An allocation method based on priority queues assigns the tasks according to their waiting times and throughput, ensuring a significant prioritization of tasks manner. The experimental results shows that proposed SC-ESOA attains the minimum energy consumption of 1.65 J and minimum delay of 1.03 s, respectively, as compared to the existing methods like ESOA. The future work will involve the hybrid optimization algorithm to enhance the overall system performance.

References

Alajlan AM, Razaque A (2023) ESOA-HGRU: Egret swarm optimization algorithm-based hybrid gated recurrent unit for classification of diabetic retinopathy. Artif Intell Rev 56(Suppl 2):1617–1646

Amer AA, Talkhan IE, Ahmed R, Ismail T (2022) An optimized collaborative scheduling algorithm for prioritized tasks with shared resources in mobile-edge and cloud computing systems. Mobile Netw Appl 27(4):1444–1460

Chai F, Zhang Q, Yao H, Xin X, Gao R, Guizani M (2023) Joint multi-task offloading and resource allocation for mobile edge computing systems in satellite IoT. IEEE Trans Veh Technol 72(6):7783–7795

Chen H, Deng S, Zhu H, Zhao H, Jiang R, Dustdar S, Zomaya AY (2022a) Mobility-aware offloading and resource allocation for distributed services collaboration. IEEE Trans Parallel Distrib Syst 33(10):2428–2443

Chen Z, Francis A, Li S, Liao B, Xiao D, Ha TT, Li J, Ding L, Cao X (2022b) Egret swarm optimization algorithm: an evolutionary computation approach for model free optimization. Biomimetics 7(4):144

Goudarzi S, Soleymani SA, Wang W, Xiao P (2023) Uav-enabled mobile edge computing for resource allocation using cooperative evolutionary computation. IEEE Trans Aerosp Electron Syst 59(5):5134–5147

Jiang H, Dai X, Xiao Z, Iyengar A (2022) Joint task offloading and resource allocation for energy-constrained mobile edge computing. IEEE Trans Mob Comput 22(7):4000–4015

Liu J, Guo S, Wang Q, Pan C, Yang L (2023) Optimal multi-user offloading with resources allocation in mobile edge cloud computing. Comput Netw 221:109522

Liu Q, Mo R, Xu X, Ma X (2024) Multi-objective resource allocation in mobile edge computing using PAES for Internet of Things. Wirel Netw 30(5):3533–3545

Mahmood OA, Abdellah AR, Muthanna A, Koucheryavy A (2022) Distributed edge computing for resource allocation in smart cities based on the IoT. Information 13(7):328

Tilahun FD, Abebe AT, Kang CG (2022) DRL-based distributed resource allocation for edge computing in cell-free massive MIMO network. In: GLOBECOM 2022–2022 IEEE global communications conference. IEEE, Rio de Janeiro, Brazil, pp 3845–3850

Wang T, Lu B, Wang W, Wei W, Yuan X, Li J (2022a) Reinforcement learning-based optimization for mobile edge computing scheduling game. IEEE Trans Emerg Top Comput Intell 7(1):55–64
Wang Z, Lv T, Chang Z (2022b) Computation offloading and resource allocation based on distributed deep learning and software defined mobile edge computing. Comput Netw 205:108732
Yang J, Wang Y, Li Z (2022) Inverse order based optimization method for task offloading and resource allocation in mobile edge computing. Appl Soft Comput 116:108361
Zhou A, Li S, Ma X, Wang S (2022) Service-oriented resource allocation for blockchain-empowered mobile edge computing. IEEE J Sel Areas Commun 40(12):3391–3404

Chapter 4
Trust and Energy-Efficient Routing Using Density Factor of Quasi-Cosine-Honey Badger Algorithm in Edge-Fog Computing

B. Sravankumar

Abstract Wireless sensor networks (WSNs) have a limited battery powered sensor which are spatially dispersed and endlessly gather data associated with the nearby environments. Edge and fog computing provide better bandwidth efficiency than cloud computing due to it process data outside the cloud which results in minimal bandwidth. However, choosing reliable and energy-efficient paths is challenging while ensuring secure communication which results in suboptimal performance. This research proposes Density Factor of Quasi-Cosine-Honey Badger Algorithm (DFQC-HBA) for trust and energy-efficient clustering and routing in WSN using edge-fog computing. The DFQC-HBA optimize routing by considering node density which provides energy efficient. The proposed DFQC-HBA increase network performance by avoiding malicious nodes which enhance both reliability and throughput. The trust, intra-cluster distance, and node degree are used as a fitness function to select secure cluster head (SCH). In routing, the distance and energy are considered to assist in optimizing path selection which minimize energy consumption and delays. Hence, the proposed DFQC-HBA achieves a less energy consumption of 17.65 mJ for 200 rounds compared to Sunflower Optimization Approach (SOA) and Firefly Approach (FA), respectively.

Keywords Density factor of Quasi-Cosine-Honey Badger algorithm · Edge-fog computing · Energy efficient · Routing · Secure cluster head · Wireless sensor networks

---

B. Sravankumar (✉)
Department of Computer Science and Engineering (AI&ML), Vaagdevi College of Engineering, Warangal, India
e-mail: sravan.researcher@gmail.com

G. P. Ramesh et al. (eds.), *Empowering IoT with Distributed Edge Computing*, IoT and Wireless Communication Networks, https://doi.org/10.1007/978-981-96-9455-6_4

Introduction

WSN contains multitudinous sensor nodes (SN) which are low-power, low-priced, and miniature characteristics. These nodes are datacentric which is responsible for gathering appropriate data in target area and transferring data to a sink node in multi or single-hop manner (Han et al. 2022). Thus, an appropriate energy-aware routing mechanism was essential for balancing the traffic load to preserve the nodes which enhance the maximum lifetime coverage in WSN. The sensor network and WSN are used in the network with physical security and limited bandwidth to minimize the energy utilization (Srinivasiah et al. 2023). Fog computing considered to be a clustered computing structure which denotes toward utilizing a cloud computing to an edge organization network. An edge network is established on a numerous interconnected device which is crucial for computation, communication, and IoT storage applications. Fog computing is an extension of cloud computing and contains primarily cloud computing characteristics like storage, encoding, networking, and computing (Moussa et al. 2022). Edge nodes are primarily placed among centralized cloud server and device by the system of comprehensive communication management (Jalasri and Lakshmanan 2023). WSN lacks energy efficiency that is a vital drawback to manage network lifetime which affects the timely data deliveries. WSN infrastructure is unique and different from conventional networks (Gurram et al. 2022). Hence, routing protocol employs SN effectively that is significant to increase network lifetime and thus, energy conservations are organized in WSN (Vellaichamy et al. 2023). In Internet Protocol version 6 (IPv6), the Routing Over Low power Lossy Network (RoLL) is a typical routing system over Low power Wireless Personal Area Network (6LoWPAN) network (Rajeesh Kumar et al. 2023). The security mechanisms by applying trust have been established to make WSN to process in a secure environment by solving an internal attack on WSN (Malik et al. 2023). Therefore, a trust-based security mechanism is used to determine the node behavior depending on historical behavior (Pathak et al. 2022).

The main contribution of this research is discussed below:

- In conventional HBA, the DFQC is included in exploration phase to avoid local optima problem and enhance convergence speed. This leads to more robust process which ensures the model to determine better global solution.
- The intra-cluster distance, trust, residual energy, distance, and intra-cluster distance are applied as a fitness function for secure clustering and routing process.
- Hence, the proposed DFQC-HBA increase security by prioritizing energy efficient and routing paths effectively.

The remaining portion is illustrated as: Section "Literature Survey" describes a literature survey of existing methods; Section "Proposed Methodology" presents an brief explanation of proposed methodology; Section "Experimental Results" evaluates an experimental results; and Section "Conclusion" presents a conclusion of a paper.

Literature Survey

Wang et al. (2024) established a two-way trust management system (TMS) to enhance security in the environment of fog computing. The TMS enhances reliability, minimize energy consumption, reduce cost, and high resource availability in cloud-fog by evaluating examination of trust-based routing method. This system integrated both Quality of Service (QoS) parameters and social trust data to compute a trust levels of fog nodes by recommendation and direct observation. The TMS enhances dependability and reduce program execution time by enhancing user satisfaction and service provider in cloud. Nevertheless, the TMS faces challenges in scalability issue because of increased overhead while managing trust among numerous nodes and devices.

Bakhtiari et al. (2024) presented a genetic algorithm (GA) and learning automation (LA) in fog computing for managing trust. The presented approach efficiently searches for trusted nodes in a high solution space and determine the best solution for the trust management optimization issue. The GA and LA adjust dynamic trust factors depending on feedback from environmental scenarios. The presented approach determined various trust factors to assign weights which achieves optimal balance among security and performance. However, the LA struggled with adaptability in trust variations because of its pre-defined learning rules which limits heterogenous and transient nature of fog nodes.

Zhang et al. (2024) developed an energy-aware two-way trust routing (ETWTR) for energy efficient and trust-based system in fog computing. The developed approach accomplished routing performance to enhance data transmission, energy availability, and reliability. ETWTR considered security, QoS, and consumption of energy simultaneously to solve the constraints of fog computing. The developed ETWTR approach enable both service provider and requesters to checked each other's reliability for enhancing security. Nonetheless, the ETWTR was reliance on frequent trust updates which results in increased latency in dynamic network conditions.

Ali et al. (2024) established a fog-cloud model for a distributed trust service which makes WSN to determine trusted mobile components to aggregate the data. The established approach generates WSN with a flexibility for aligning with a trust policy based on data aggregation process which provides effective access to trust service. The distributed trust mechanism allows WSN to process various aspects based on threshold values. The global and local trust were integrated for assigning the values of final trust to mobile elements (MEs) in the established system. However, bandwidth usage and latency were enhanced while aggregating data from mobile elements because frequent communication among cloud and fog nodes.

Akbari et al. (2022) introduced an overlapping clustering Method in Fog Computing Technology in Internet of Things (MFCT-IoT) to choose an optimal cluster head (CH) nodes for providing the rapid data transfer from objects to fog nodes. A chosen CH node were responsible for transmitting gathered data to nearby fog in network edge. An introduced approach minimizes packet loss rate by organizing a fog node in hierarchical tree to transmit data effectively in cloud. Nevertheless, the introduced approach was inefficient in resource allocation because multiple CH leads to redundant data storage and processing.

Proposed Methodology

This research proposes DFQC-HBA for energy efficient and trust-aware clustering and routing in WSN based on edge-fog computing. The DFQC-HBA selects SCH and routing path and organizes clusters effectively. The intra-cluster distance, node degree, trust (direct and indirect), residual energy, and distance are the five-fitness function considered for clustering and routing process. Then, a data is transmitted to BS once an best CH and routing path are established.

Edge-Fog Node

In an edge-fog network, the main goal of trust management system is to determine and minimize a malicious fog node and inappropriate fog clients. To minimize vagueness and predict future trust, the system extracts particular object information from direct data and recommendations. The exchange series among fog node and fog client helps for establishing a dependable connection. Then, a connection request is transmitted from client to fog for determining trust score for client. Whether a client is assumed trustworthy, access is allowed; whether not, it is denied. Based on establishment on connection, a fog's trust score is evaluated, and if considered reliable, a connection is confirmed.

Node Initialization

It involves setting up the individual SN with required configurations like assigning unique identifiers, ensuring energy-efficient process, and establishing communication parameters. Moreover, the nodes determined the roles either as SN, intermediate relay nodes, or potential edge-fog node during initialization which manages local data storage and processing. This initialization stage ensures seamless combination with fog and edge layers which provides effective data processing and minimized latency. Once the initialization is determined, then the SCH is performed using proposed approach.

SCH Using DFQC-HBA

HBA (Hashim et al. 2022) is a swarm intelligence approach and prioritizes nodes based on trustworthiness and residual energy. It dynamically adjusts routing paths to avoid malicious or low-energy nodes which ensures reliable and secure communication. The algorithm balances the consumption of energy across nodes to increase network lifetime. Hence, its adaptive nature increases routing performance in heterogenous edge-fog environments. A detailed information about the proposed DFQC-HBA is represented in Fig. 4.1.

Characterizing the Intensity (*I*). A moderation represents a strength and distance among search and target agent. A search agent movement is majorly determined by its factory cues and a high fragrance intensity leads to rapid motion. A mathematical formula for fragrance intensity is expressed in Eqs. (4.1)–(4.3)

$$I_i = r_2 \times \frac{S}{4\pi d_i^2} \tag{4.1}$$

$$S = (x_i - x_{i+1})^2 \tag{4.2}$$

$$d_i = x_{\text{prey}} - x_i \tag{4.3}$$

where I_i indicates a target's fragrance, x_i represents ith search space, S denotes concentration strength, d_i determines gap among target and ith search member, and x_{prey} illustrates target's finest place.

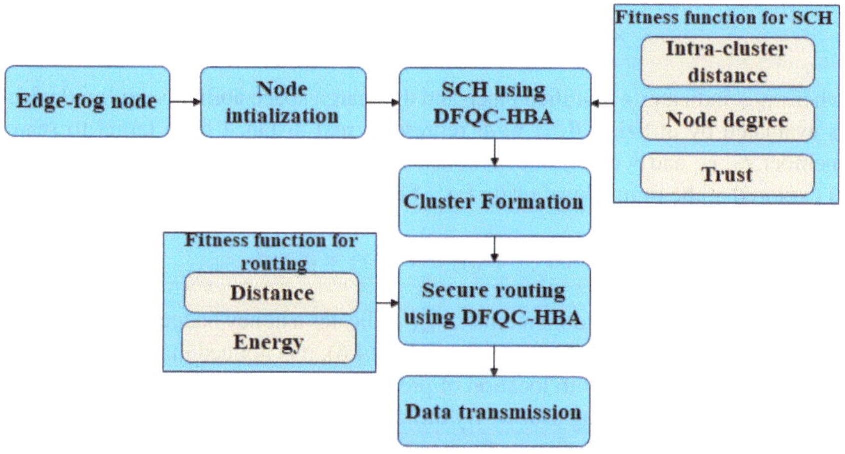

Fig. 4.1 Workflow of the proposed methodology

Modify Density Factor. A density factor α ensures a HBA's smooth transition from global exploration to local deployment. Moreover, α iteratively minimize by period which is represented using Eq. (4.4)

$$\alpha = C \times \exp\left(\frac{-t}{t_{\max}}\right) \tag{4.4}$$

where $C = 2$ represents constant and $t_{\max}$ indicates maximum number of iterations. The density factor is a significant component to balance the exploitation and exploration phase. In the Eq. (4.4), the α minimizes continuously in the concave function shape with increase in number of iterations which enhance the global optimization process. Nevertheless, the alteration of α is gentle that minimize the model's convergence speed. Moreover, in the middle of model iteration, and decrease continuously in the density factor results in local optima issue. Hence, the DFQC is applied in the HBA using Eqs. (4.5) and (4.6)

$$\alpha = |1.7 \times \omega \times \cos(2\pi + 1.3 \times \cos(2\pi \cdot \omega))| \tag{4.5}$$

$$\omega = \cos\left(\frac{-t}{t_{\max} \cdot 2} \times \pi\right) + 0.2 \tag{4.6}$$

where t determines present iteration number and $t_{\max}$ indicates a maximum iteration number.

Digging and Honey Phase. A honey badger acts as correspondingly to Cardioid shape curve using Eq. (4.7)

$$\begin{aligned} x_{\text{new}} = x_{i-\text{prey}} + F \times \beta \times I \times x_{\text{prey}} + F \times r_3 \times \alpha \times d_i \\ \times |\cos(2\pi r_4) \times [1 - \cos(2\pi r_5)]| \end{aligned} \tag{4.7}$$

where x_{new} indicates a modified place and the search space ability is explored which is evaluated by utilizing $\beta = 6$, d_i represents real distance from target to search member, r_3, r_4, and r_5 determine random number. In honey phase, the search member is explored at the mimic level using Eq. (4.8)

$$x_{\text{new}} = x_{i-\text{prey}} + F \times r_7 \times \alpha \times d_i \tag{4.8}$$

where r_7 determines a random number, x_{new} indicates a honey badger's new position, and $x_{i-\text{prey}}$ represents prey location. From Eq. (4.6), it is noticed that a honey badger generates a search nearby to location of prey. In this phase, search is determined by search behavior by varying time α. By utilizing DFQC in HBA increase node selection by better balancing energy efficiency and trust. It improves decision-making process for spatial and density node distribution that optimize routing paths. This factor minimizes energy consumption by reducing unnecessary node while maintaining secure communication. Moreover, it ensures even energy utilization among nodes which increase network lifetime.

Fitness Function

The fitness function like intra-cluster distance, node degree, trust (direct and indirect), residual energy, and distance are employed for secure clustering and routing in edge-fog computing. The discussion about this fitness function is explained below in detail.

Intra-Cluster Distance. It is a Euclidean distance among non-mobile SN to its associated mobile node. A SN generates high energy in data transmission process and whether a transmission distance is less, then less energy is need to generate a data using Eq. (4.9)

$$F_1 = \sum_{j=1}^{m} \left(\sum_{i=1}^{I_j} \mathrm{dis}(s_i, \mathrm{CH}_j) / I_j \right) \tag{4.9}$$

where F_1 indicates a fitness function of intra-cluster distance, m denotes total number of clusters, I_j determines number of SN in jth cluster, s_i represents ith SN in jth cluster, and CH_j evaluates distance among ith SN in jth cluster.

Node Degree. It is an amount of non-CH members which is belongs to its associated mobile node and whether a mobile has involved a smaller number of members, then it sustains for long duration using Eq. (4.10)

$$F_2 = \sum_{i=1}^{m} I_i \tag{4.10}$$

Trust. Direct trust (DT) is computed among node and node at time using Eq. (4.11). Indirect trust (IDR) employs a social factor generated by neighboring nodes to increase trustworthiness accuracy which is represented using Eq. (4.12). The overall trust function is expressed using Eq. (4.13)

$$T_{i,j}^{\mathrm{direct}}(t) = \sum_{p=1}^{q} \mathrm{d}z_p Q_p \tag{4.11}$$

$$T_{i,j}^{\mathrm{indirect}}(t) = \frac{\sum_{k=1}^{n} T_{i,j}^{\mathrm{direct}}(t) \times (z_1 \times T_{\mathrm{FRI}}(t)) \times (z_2 \times T_{\mathrm{HY}}(t))}{n} \tag{4.12}$$

$$F_3 = W_1 T_{i,j}^{\text{direct}}(t) + W_2 T_{i,j}^{\text{indirect}}(t) \tag{4.13}$$

where dz_p determines assigned weight, p indicates minimum QoS parameter, q illustrates number of QoS parameter, z_1 and z_2 represent weight, n determines total number of interactions, k illustrates index, $T_{\text{FRI}}(t)$ denotes trust element depending on interaction frequency at time t, W_1 and W_2 indicate a weight factor of $T_{i,j}^{\text{direct}}(t)$ and $T_{i,j}^{\text{indirect}}(t)$, such that $W_1 + W_2 = 1$, and, $T_{\text{HY}}(t)$ show trust element based on metrics.

Energy. Mobile node obtains data from non-CH nodes in the data transmission stage and then transmits to event position after aggregation. Mobile nodes need more residual energy to generate these tasks using Eq. (4.14)

$$F_4 = \sum_{i=1}^{m} \frac{1}{E_{\text{CH}_i}} \tag{4.14}$$

Distance. It is a Euclidean distance among each mobile node to event position using Eq. (4.15)

$$F_5 = \sum_{i=1}^{m} \left(\text{dis}\left(\text{CH}_j, \text{EL}\right)\right) \tag{4.15}$$

The overall fitness function formula F is represented using Eq. (4.16)

$$F = \rho_1 \times F_1 + \rho_2 \times F_2 + \rho_3 \times F_3 + \rho_4 \times F_4 + \rho_5 \times F_5 \tag{4.16}$$

where $\rho_1, \rho_2, \rho_3, \rho_4, \rho_5, \rho_6, \rho_7$ represents weighting factor.

Cluster Formation

In WSN, it involves grouping SN into clusters which is handled by a CH. In edge-fog computing, cluster formation is essential for increasing scalability and processing effectiveness by enabling localized data processing at fog nodes. It minimizes latency, increase network lifetime, and optimize resource utilization by balancing energy consumption between nodes. Moreover, it also increases data reliability and supports hierarchical routing in dynamic and resource-constrained edge-fog environments.

Secure Routing Process

Secure routing in WSN ensures confidentiality, data security, and reliability in edge-fog computing while optimizing energy usage. The process contains authenticating nodes which selects trust-based and energy-efficient routes. Fog nodes provide localized processing data, minimize latency from the edge-fog. Trust models determine node behavior to identify malicious nodes in routing paths. The DFQC-HBA leverages spatial distribution and node density for optimized path selection. The density factor makes the model prioritize trusted nodes with appropriate residual energy which improves reliability and security. The quasi cosine law variation enhances path adaptability by considering proximity and angular relationships among nodes. Hence, this model reduces routing by malicious or compromised nodes though balancing energy consumption. It increases network lifetime by avoiding energy depletion in densely associated areas. Therefore, it ensures secure, energy-efficient, and adaptive routing in WSNs for edge-fog computing environments.

Experimental Results

The DFQC-HBA is simulated using MATLAB R2021b with windows 10 operating system, 64 GB RAM, and an i5 intel processor. The consumption of energy, delay, and throughput are the performance measures applied to evaluate a model performance in this research. Table 4.1 determines a simulation parameter.

Performance Measures

Table 4.2 represents an analysis of energy consumption (mJ). The existing techniques like SOA, FA, and HBA are compared with proposed DFQC-HBA method. While compared to these techniques, the proposed DFQC-HBA obtains a less energy consumption of 17.65 mJ by optimizing node selection depending on angular and density positioning. It prevents routing by distant nodes which minimize inappropriate transmission of data. The quasi-cosine law makes effective path adaptation to reduce the number of active nodes for communication. By balancing the utilization of energy over nodes, it avoids early depletion in significant areas in edge-fog environment.

Table 4.1 Simulation parameter

Parameters	Values
Energy of fog server	10 J
Bandwidth	54 Mbps
Storage of fog server	5 GB
Iterations	100
Number of nodes	100

Table 4.2 Analysis of energy consumption (mJ)

Methods	No. of nodes	No. of rounds				
		200	400	600	800	2000
SOA	100	39.67	43.50	47.79	52.58	59.67
FA		32.58	39.67	41.37	47.69	52.48
HBA		22.47	26.89	32.58	38.56	43.79
DFQC-HBA		17.65	21.48	26.09	28.48	32.47

Table 4.3 indicates a performance analysis of delay (s). The proposed DFQC-HBA achieves a less delay of 0.34 s by optimizing routing paths which considers density of node distributions. It avoids low-trust nodes by minimizing retransmission and processing delay. The quasi-cosine law makes effective path selection by reducing angular deviations which results in shorter routes. Hence, the adaptive routing mechanism makes effective data transmission in edge-fog compared to existing methods like SOA, FA, and HBA.

Table 4.4 illustrates a performance analysis of throughput. A DFQC-HBA achieves a better throughput of 99.76% compared to SOA, FA, and HBA due to the proposed method's effective node selection and routing progress. The density factor makes balanced traffic distribution which avoids congestion in associated areas. The quasi-cosine law variation increases path adaptability by prioritizing nodes depending on angular efficiency and proximity which provide reliable data transmission. Moreover, its energy-efficient routing reduces node failures and maintains network stability and sustains high throughput across time.

Table 4.3 Analysis of delay (s)

Methods	No. of nodes	No. of rounds				
		200	400	600	800	2000
SOA	100	1.67	1.92	2.37	2.83	2.98
FA		0.98	1.45	1.87	1.92	2.34
HBA		0.67	0.87	1.36	1.76	1.98
DFQC-HBA		0.34	0.42	0.53	0.76	1.34

Table 4.4 Analysis of throughput (%)

Methods	No. of nodes	No. of rounds				
		200	400	600	800	2000
SOA	100	91.25	90.35	89.64	87.26	86.33
FA		92.47	92.13	97.36	97.14	96.58
HBA		97.29	96.87	96.54	95.87	95.47
DFQC-HBA		99.76	99.54	98.76	98.32	97.01

Discussion

An advantage of proposed DFQC-HBA and limitations of existing techniques are discussed in this section in detail. A limitation of existing methods like TMS (Wang et al. 2024) faces challenges in scalability issue because of increased overhead while managing trust among numerous nodes and devices. LA (Bakhtiari et al. 2024) struggled with adaptability in trust variations because of its pre-defined learning rules which limits heterogenous and transient nature of fog nodes. ETWTR (Zhang et al. 2024) was reliance on frequent trust updates which results in increased latency in dynamic network conditions. Bandwidth usage and latency in were enhanced in fog-cloud model (Ali et al. 2024) while aggregating data from mobile elements because frequent communication among cloud and fog nodes. The proposed DFQC-HBA overcomes these existing method limitations by balancing energy consumption through prioritizing nodes with optimal density which increase network lifetime. The quasi-cosine law variation increase routing by processing secure and effective paths which minimize communication delays. Its trust-aware model reduces malicious attacks by avoiding untrustworthiness which enhance security and data integrity. Moreover, the localized processing in edge-fog minimize delays and generates rapid and reliable decision-making process and this ensures a robust and energy-efficient WSN.

Conclusion

This research proposes DFQC-HBA for trust and energy-efficient clustering and routing in WSN based on edge-fog environment. In HBA, the DFQC is incorporated to avoid local optima issue and enhance convergence speed. The proposed approach determines nodes depending on the density and angular effectiveness which enhance path adaptability. This makes secure, reliable communication, and energy efficient by preventing from untrustworthy nodes. The intra-cluster distance, trust, distance, node degree, and residual energy are used as a fitness function which ensures secure and effective clustering and routing. The CH selection and route path discovery are performed by using proposed DFQC-HBA. The trust avoids malicious nodes and enhance network security hence the combination of DFQC-HBA improves overall network performance in WSN based on edge-fog computing. Compared to existing methods such as SOA and FA, the proposed DFQC-HBA achieves a less energy consumption of 17.65 mJ for 200 rounds.

References

Akbari MR, Barati H, Barati A (2022) An efficient gray system theory-based routing protocol for energy consumption management in the Internet of Things using fog and cloud computing. Computing 104(6):1307–1335

Ali BA, Abdulsalam HM, Almonaies A, Alroumi E (2024) A cloud-fog distributed trust service for wireless sensor networks. J Supercomput 80(16):24578–24604

Bakhtiari NB, Rafighi M, Ahsan R (2024) A trust management system for fog computing using improved genetic algorithm. J Supercomput 80:20923–20955

Gurram GV, Shariff NC, Biradar RL (2022) A secure energy aware meta-heuristic routing protocol (SEAMHR) for sustainable IoT-wireless sensor network (WSN). Theoret Comput Sci 930:63–76

Han Y, Hu H, Guo Y (2022) Energy-aware and trust-based secure routing protocol for wireless sensor networks using adaptive genetic algorithm. IEEE Access 10:11538–11550

Hashim FA, Houssein EH, Hussain K, Mabrouk MS, Al-Atabany W (2022) Honey Badger algorithm: new metaheuristic algorithm for solving optimization problems. Math Comput Simul 192:84–110

Jalasri M, Lakshmanan L (2023) Managing data security in fog computing in IoT devices using noise framework encryption with power probabilistic clustering algorithm. Clust Comput 26(1):823–836

Malik TS, Tanveer J, Anwar S, Mufti MR, Afzal H, Kim A (2023) An efficient and secure fog based routing mechanism in IoT network. Mathematics 11(17):3652

Moussa N, El Belrhiti El Alaoui AL (2022) DACOR: a distributed ACO-based routing protocol for mitigating the hot spot problem in fog-enabled WSN architecture. Int J Commun Syst 35(1):e5008

Pathak A, Al-Anbagi I, Hamilton HJ (2022) An adaptive QoS and trust-based lightweight secure routing algorithm for WSNs. IEEE Internet Things J 9(23):23826–23840

Rajeesh Kumar NV, Jaya Lakshmi N, Mallala B, Jadhav V (2023) Secure trust aware multi-objective routing protocol based on battle competitive swarm optimization in IoT. Artif Intell Rev 56(Suppl 2):1685–1709

Srinivasiah VPB, Ranganathasharma RH, Ramanna V (2023) TCRP: trust-aware clustering and routing protocol based on atom search optimization for WSNs. Int J Intell Eng Syst 16(4):581–590

Vellaichamy J, Basheer S, Bai PSM, Khan M, Kumar Mathivanan S, Jayagopal P, Babu JC (2023) Wireless sensor networks based on multi-criteria clustering and optimal bio-inspired algorithm for energy-efficient routing. Appl Sci 13(5):2801

Wang J, Luo Z, Wang C (2024) A two-way trust routing scheme to improve security in fog computing environment. Clust Comput 27:13165–13185

Zhang Y, Yu Y, Sun W, Cao Z (2024) Towards an energy-aware two-way trust routing scheme in fog computing environments. Telecommun Syst 87:973–989

Chapter 5
Multi-agent Deep Reinforcement Learning with Stochastic Gradient Descent for Peer-to-Peer Computation Offloading in IoT Edge Computing

K. V. Sheelavathy, V. Prabhudeva, D. Kannan, and Debashis Rudra Sarma

Abstract Peer-to-peer (P2P) computation offloading has an Internet of Things (IoT) device to nearby device with accessible resources. This process involves a device-to-device (D2D) communication and minimize dependency on centralized cloud servers and enhance the effectiveness of task execution. P2P offloading increase resource utilization by sharing the workload over numerous devices which minimize energy consumption. However, the inefficient allocation of computational tasks over IoT device cause high delay and energy because of lack of coordination and task distribution. In this research, the multi-agent deep reinforcement learning with stochastic gradient descent (MADRL-SGD) is proposed for P2P Computation Offloading in IoT Edge Computing. It makes effective and dynamic tasks allocation by applying multi-agent collaboration which helps to minimize energy and latency. The SGD provides rapid convergence and enhanced learning effectiveness in IoT edge computing. MADRL adapts to heterogenous device by varying workloads and capabilities which increase overall system performance. The proposed MADRL-SGD achieves a low average total cost of 0.5 for task arrival probability compared to existing methods like deep Q-network (DQN) and DRL, respectively.

Keywords Device-to-device · Edge computing · Internet of Things · Multi-agent deep reinforcement learning · Peer-to-peer · Stochastic gradient descent

---

K. V. Sheelavathy
School of Computer Science and Engineering, Reva University, Bengaluru, India
e-mail: Sheela.kv@reva.edu.in

V. Prabhudeva
Enterprise Cloud Migration Team, Tata Consultancy Services Ltd, Bengaluru, India
e-mail: prabhu.v@tcs.com

D. Kannan
Lead Cloud PaaS Engineering, TCS, Think Campus, Bengaluru, India

D. R. Sarma (✉)
Data and Analytics Group, Life Science Health Care, Tata Consultancy Services, Bengaluru, India
e-mail: Debashis.sarma@tcs.com

G. P. Ramesh et al. (eds.), *Empowering IoT with Distributed Edge Computing*, IoT and Wireless Communication Networks, https://doi.org/10.1007/978-981-96-9455-6_5

Introduction

Internet has converted from mobile-Internet and peer-to-peer (P2P) networking in the last few years to the world wide web and Internet of Things (IoT) that receive huge volume of data. However, the single device does not have sufficient capacity to evaluate the data by utilizing a cloud computing. The deployment of remote clouds does not satisfy the requirements of various services and IoT applications (Almashhadani et al. 2022). Furthermore, cloud computing involves a longer waiting time compared to edge computing. Therefore, edge computing is applied to minimize the bandwidth consumption and data transfer delay (Zeng et al. 2023). Edge computing forces abundant computation resources such as memory, central processing unit (CPU), and storage from a centralized cloud to edge which makes densely applied edge servers to generate computation services for the application of intensive computing (Robles-Enciso and Skarmeta 2023). The application like P2P computation offloading is segregated into various tasks. Based on this, task offloading with fine-granularity in edge computing is studied extensively where the tasks are offloaded from user device to edge servers for energy consumption and low application delay (Dai et al. 2023) (Tang et al. 2022). An edge computing is established for managing user connectivity in same neighborhood to create an autonomous mobile peer-to-peer streaming overlay network (MPPS). It contains a sliding window cache which is associated with each peer for establishing a video segment (Almashhadani et al. 2023). A peer initially checks with MPPS server for absent packets in sliding window to determine peers who supplied he absent packets (Ren et al. 2024). Then, the peer gives the request from one to other peers to transmit the packets. The peer goes to the cloud servers for downloading the packets if no peers supply the packets (Jalilvand Aghdam Bonab and Shaghaghi Kandovan 2022). Meanwhile, each peer uses its uplink for uploading data because the more peers contribute in a MPPS leads to more traffic in a backhaul network (Tong and Yang 2022).

The main contribution of this research is discussed below in detail,

- MADRL makes collaborative and dynamic decision-making multiple agents for effective computation offloading. Moreover, the MADRL scales in intricate and large IoT networks which provides better utilization.
- SGD generates stable and rapid convergence in the training of DRL and enhance learning effectiveness in uncertain and dynamic IoT edge computing.
- Hence, the combination of MADRL and SGD reduces task delay and enhances resource utilization effectively.

The remaining portion is represented as: Section "Literature Survey" determines the literature survey of existing techniques; Section "Proposed Methodology" explains a detailed discussion of proposed methodology; Section "Experimental Results" evaluates the experimental results; and Section "Conclusion" shows a conclusion of the research.

Literature Survey

The related work about P2P computation offloading was discussed in detail along with its advantages and limitations.

Mu and Shen (2022) presented a stochastic learning for opportunistic P2P offloading in IoT edge computing. Each requestor enables decisions on both offloading power of transmission and local computation frequency to reduce its task completion cost by considering task delay, energy consumption, and task loss because of buffer overflow. The dynamic decision process between several requestors was determined as a stochastic game. A decentralized online offloading was utilized by establishing a post-decision state where each requestor as an independent agent learning for determining optimal strategies. However, the stochastic learning was inefficiency in managing dynamic network conditions and unpredictable user behavior which results in suboptimal offloading decisions.

Zhang et al. (2023) established a satellite peer offloading mechanism where offloading is established along multi-hop paths which explores collaborative computing abilities. Then, the Multi-hop Satellite Peer Offloading (MHSPO) issue was determined which helps to reduce the energy consumption and delay under backlog constraints and system resources. At last, the practical online distributed approach was used to address the MHSPO which obtains a close-to-optimal performance. Nevertheless, the MHSPO suffers from network reliability problem because multi-hop links were prone to disruptions in dynamic environments.

Chi and Radwan (2022) introduced a fully decentralized fairness aware federated mobile edge computing peer offloading for enterprise management network. A federated gradient descent with fully decentralized approach was established into separate edge computing-small cell which obtains a global optimum without either aggregation of data or service provider sharing. The performance of introduced approach was determined by a comprehensive simulation set that obtains a high convergence with latency, Quality of Service (QoS), and load balancing. However, the fully decentralized approach has an unequal resource allocation because of varying device abilities which leads to reduced distribution of task.

Zhou et al. (2023) suggested a deep reinforcement learning-based computation offloading and service caching method (DRLCOSCM) to enhance the service caching, offloading decision, and resource allocation strategies by ensuring mobile user's delay requirements. In DRLCOSCM, the optimization issue was determined as an Asynchronous Advantage Actor-Critic (A3C) and Mixed Integer Non-Linear Programming (MINLP) to address the issue of computation offloading. Nevertheless, the DRLCOSCM struggled to adapt to rapidly changing network environments because of dependence on pre-trained approach.

Liu et al. (2023) developed an Asynchronous update Reinforcement Learning-based Offloading (ARLO) for mobile edge computing. Every sub-network involved a same structure because it interacts with environment for learning and updating the public network. The sub-network determined the central public network parameters effectively. The primary aim for utilizing asynchronous multithreading was that it enables threads for learning the strategy at same time which makes learning process rapidly. However, the ARLO suffers from non-convergent updates because of delayed gradient propagation which leads to minimized task scheduling.

From the overall analysis, the existing methods had limitations such as inefficiency in managing dynamic network conditions and unpredictable user behavior, suffers from network reliability, unequal resource allocation, and difficulties in adapting to rapid changing network environments. To address these issues, the MADRL with SGD is proposed for P2P computation offloading in IoT edge computing. This approach optimizes decision-making by making adaptive and dynamic allocation. Moreover, this approach increases network reliability via continuous learning over various agents. The deployment of SGD makes rapid convergence and enhanced effectiveness in P2P computation offloading.

Proposed Methodology

In this research, the MADRL with SGD is proposed for P2P computation offloading in IoT edge computing. MADRL is included to enable decentralized and adaptive computation offloading. It allows IoT devices for optimizing task offloading strategies by considering interactions with other devices in a stochastic environment.

System Model

Assume a edge computing network involves a number of IoT devices and certain devices called requestors which are loaded heavily with computation intensive task. An requestor offloads part of its tasks to a helper using D2D link. Then, a helper utilizes its computing resources to execute an offloaded task whenever it does not process its tasks. Before the offloading process, the formation of Requestor-Helper (RH) pair and link establishment progress are performed. Whether a number of requestors is greater than number of helpers, a rest of requestors does not able to determine a helper to progress a task. These requestors do not benefit from P2P offloading and hence are not measured in a below design. Assume $f_m^L(t)$ represents CPU requestor frequency m at time slot t that is called internal action managed by requestor m. The local computation energy disbursed by m at time t is represented using Eq. (5.1)

$$E_m^L(t) = k\left(f_m^L(t)\right)^3 T \tag{5.1}$$

Equation (5.2) shows a number of tasks which are executed locally by requestor m at t.

$$T_m^L(t) = \lfloor f_m^L(t)\tau / L_m \rfloor \tag{5.2}$$

where $\lfloor . \rfloor$ indicates a rounding down to integer. Assume that each RH pairs shared the equivalent transmission of wireless channel with bandwidth W for computation offloading process. During one time slot, the channel state remains constant however change over various time slot. Based on correlated fading channel among two devices is determined by the finite state Markov chain in state k which is represented as $k = \{1, 2, \ldots, K\}$. The possibility of transmission among 2 nearby states k and $k+1$ is represented as $P_{k,k+1}$, $1 \le k < k$. Based on Shannon formula, the transmission rate among helper and requestor of RH pair m at t is determined in Eq. (5.3)

$$R_m(t) = W \log\left(1 + \frac{p_m(t) H_{\mathrm{mm}}(t)}{\sigma^2 + \sum_{i \in M/\{m\}} p_i(t) H_{\mathrm{im}}(t)}\right) \tag{5.3}$$

where $H_{\mathrm{im}}(t)$ represents channel gain among helper m and requestor i, σ^2 indicates background power of white Gaussian noise. The energy consumption acquired by requestor m to transmit an offloaded task's input data which is expressed using Eq. (5.4)

$$E_m^O(t) = p_m(t)\tau \tag{5.4}$$

To determine the task offloading performance, an instant cost function is described for requestor m at t contains an energy cost, immediate task loss cost, and delay. An instant energy cost $C^E(E_m(t))$ is an increasing energy consumption function $E_m(t)$ that has a local and transmission energy which is expressed using Eq. (5.5)

$$E_m(t) = E_m^L(t) + E_m^O(t) \tag{5.5}$$

The function of instant overall cost for requestor m at t is determined using Eq. (5.6)

$$U_m(t) = C_m^E(E_m(t)) + C_m^D(D_m(t)) + C_m^O(O_m(t)) \tag{5.6}$$

where $C_m^O(O_m(t))$ represents increasing function of dropped task $O_m(t)$.

Multi-agent Deep Reinforcement Learning (MADRL) with Stochastic Gradient Descent (SGD)

By using MADRL with SGD provides an adaptability and scalability for P2P computation offloading in IoT dynamic environments. MADRL makes decentralized decision-making which allows IoT devices for optimizing offloading strategies depending on local observations by considering interactions with another device.

Integrating SGD makes effective training for value function even with large state-action space. This combination manages stochastic components such as channel variability, task arrivals, and helper availability. It enhances resource utilization, minimize energy consumption and delays. The environment is considered as randomness and uncertainty which is influenced by probabilistic factors. Stochastic learning assists the agents for adapting to these uncertain and varying conditions by utilizing stochastic approximation which iteratively enhance the strategies. DRL (Fang et al. 2022) employs a deep neural network (DNN) for simulating the optimizer of conventional RL technique and acquires the best action decision policy π_k^* by training the DNN continuously using Eq. (5.7)

$$\pi_k^* = \underset{a_n \sim \pi_n}{\arg\max} E\left[\sum_{t=1}^{T} \gamma^{(t-1)} u_n(t)\right] \tag{5.7}$$

DQN is primarily utilized off-policy and model-free DRL approach that is solved in certain optimization issue, but the tricky issue of DQN is overestimation issue for predictable cumulative reward. Hence, DDQN is used to solve the DQN overestimation by determining a target network those structure is similar as primary network. Other advanced technology like experience replay memory is applied to address the relation among adjacent sample training. Figure 5.1 represents a MADRL with SGD depending on DDQN. Based on present state $o_n(t)$, the agent executes some action $a_n(t)$. Then it transferred into following state $o_n\prime(t)$ and acquire immediate reward $u_n(t)$. M-network n contains a wireless node training and the output is $a_n(t)$. Moreover, the T-network structure is same as M-network structure, T-network did not contribute in training but updates every parameter of neural network regularly from M-network by the pattern of soft update. Experience replay memory is applied for saving each agent's training samples. The loss function is updated for minimizing the loss by each agent's continuous training using SGD. After completing the training process, the wireless node downloads the trained neural network weight from edge server for performing offloading decision-making policy. Also, the neural network weight is numerical data and the weight of every wireless node neural network is about numerous hundred KB in size that does not impact many transmissions overhead.

Experimental Results

The proposed MADRL with SGD is simulated using Python 3.4 environment with an Intel i7 processor and windows 10 operating system. The average total cost, task delay cost, and energy cost are evaluated to determine the model performance of proposed MADRL with SGD method.

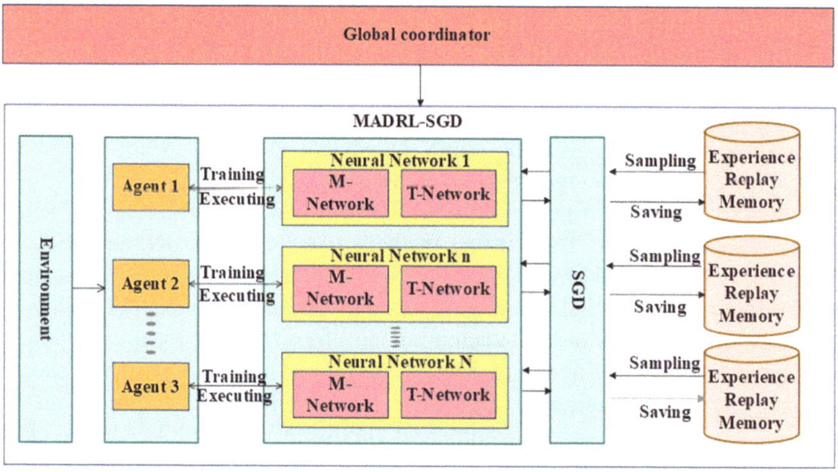

Fig. 5.1 Architecture of MADRL-SGD

Performance Analysis

Table 5.1 represents a performance analysis of task arrival probability versus average total cost. The existing techniques like DQN, Double DQN (DDQN), and DRL are compared with proposed MADRL-SGD. While compared to these existing methods, the proposed MADRL-SGD obtains a low average total cost of 0.5 due to MADRL enables IoT devices for learning optimal offloading strategies in dynamic and stochastic factors such as task arrivals. SGD ensures effective and scalable training of deep mechanism which provides rapid updates and convergence in large state-action spaces. The MADRL's decentralized nature minimize communication overhead whereas SGD enhance adaptability. Therefore, the proposed MADRL-SGD obtains a low average total cost compared to existing methods.

Table 5.2 indicates a performance analysis of task arrival probability versus average energy cost. The proposed MADRL-SGD achieves a less average energy cost of 0.2 due to it optimize offloading decisions by learning effective policies which balance both offloading and local computation. In SGD, the gradient based updates

Table 5.1 Task arrival probability versus average total cost

Methods	Task arrival probability				
	0.2	0.3	0.4	0.5	0.6
DQN	4.2	5.3	6.2	7.5	8.2
DDQN	3.1	3.5	3.6	4.5	4.9
DRL	2.1	2.6	3.4	3.6	4.5
MADRL-SGD	0.5	1.5	1.9	2	2.4

provide rapid convergence to energy-efficient solutions by reducing unnecessary usage during training. MADRL controls agent collaboration to minimize interference which results in less redundant transmission and computation. The decentralized mechanism adapts in tasks loads and peer availability which avoids in energy-intensive process. Moreover, the policy refinement via SGD avoids suboptimal actions which reduce average energy cost over time.

Table 5.3 illustrates a performance analysis of task arrival probability versus average task delay cost. The existing methods like DQN, DDQN, and DRL are compared with proposed MADRL-SGD. When compared to these existing methods, the proposed MADRL-SGD achieves a less average task delay cost of 0.5 due to it dynamically balances task allocation among offloading and local execution by employing a resource availability. The SGD updates make rapid convergence to optimal policies which minimize delay caused by suboptimal decisions. By modeling inter-agent interference, it reduces bottlenecks in shared resources like wireless channels. Moreover, the decentralized learning makes localized and effective decisions without depend on centralized coordination. This strategy makes tasks are processed with minimal transmission delays.

Table 5.4 determines a helper available probability versus average task delay cost. The existing technique like DQN, DDQN, and DRL are compared with proposed MADRL-SGD. The proposed MADRL-SGD achieves an 8.2 average task delay cost by effectively balancing offloading decisions and local processing via learned policies in dynamic system states. This approach makes effective tasks prior buffer overflow occurs. The combination of MADRL minimize delay between devices by managing offloading strategies. SGD performs a rapid convergence to optimal policies which reduce suboptimal actions during the learning process. Moreover, the decentralized

Table 5.2 Task arrival probability versus average energy cost

Methods	Task arrival probability				
	0.2	0.3	0.4	0.5	0.6
DQN	3.1	3.4	3.6	4.2	4.6
DDQN	2.0	2.4	2.8	3.4	3.7
DRL	1.2	1.5	1.8	2.4	2.8
MADRL-SGD	0.2	0.3	0.5	0.6	0.9

Table 5.3 Task arrival probability versus average task delay cost

Methods	Task arrival probability				
	0.2	0.3	0.4	0.5	0.6
DQN	2.4	2.7	3.2	3.7	5.2
DDQN	3.1	3.5	3.7	3.9	4.0
DRL	1.5	1.7	2.9	3.4	3.9
MADRL-SGD	0.5	1.2	2.5	3.4	3.6

Table 5.4 Helper available probability versus average task delay cost

Methods	Helper availability probability				
	0.2	0.3	0.4	0.5	0.6
DQN	23	16	12	9	7
DDQN	11	9	7	6.5	6.2
DRL	10.5	10	9.7	9.2	8.5
MADRL-SGD	8.2	6.5	5	4.5	4.1

model prevents bottlenecks in the execution of tasks compared to DQN, DDQN, and DRL, respectively.

Discussion

The advantage of proposed MADRL-SGD and limitations of existing techniques are discussed in detail. The limitations of existing methods such as stochastic learning (Mu and Shen 2022) were inefficiency in managing dynamic network conditions and unpredictable user behavior which results in suboptimal offloading decisions. MHSPO (Zhang et al. 2023) suffers from network reliability problem because multi-hop links were prone to disruptions in dynamic environments. Fully decentralized approach (Chi and Radwan 2022) has an unequal resource allocation because of varying device abilities which leads to reduced distribution of task. DRLCOSCM (Zhou et al. 2023) struggled to adapt to rapidly changing network environments because of dependence on pre-trained approach. ARLO (Liu et al. 2023) suffers from non-convergent updates because of delayed gradient propagation which leads to minimized task scheduling. The proposed MADRL-SGD overcomes these existing method limitations by applying its decentralized nature and adaptive learning abilities. The MADRL enable each device to learn optimal offloading strategies independently by interacting with other device which ensures effective resource utilization. By incorporating SGD, the system converges rapidly and minimize delays in decision-making and adapts to environmental uncertainties by varying network conditions. Moreover, the SGD assists in updating the model parameters effectively by providing better coordination among devices. This approach optimizes decision-making by making adaptive and dynamic allocation. Hence, the proposed MADRL-SGD reduce average energy cost and average task delay compared to the existing methods like DQN, DDQN, and DRL, respectively.

Conclusion

This research proposes MADRL-SGD for P2P computation offloading in IoT edge computing. MADRL optimizes offloading strategies based on local observations by considering interactions with another device. Stochastic learning helps the agents for adapting to these uncertain and varying conditions by applying stochastic approximation that iteratively increase the strategies. By integrating MADRL and SGL, the model makes decentralized for effective tasks offloading. MADRL enable devices for learning optimal strategies depending on local conditions whereas SGD provides rapid convergence which enhance model performance. The system adapts to dynamic environments with varying arrival of task, network condition, and resource availability. Hence, the proposed MADRL-SGD minimize average task delay, enhance resource utilization and makes better task completion in an IoT edge environment. By executing this process, the proposed MADRL-SGD obtains a less average total cost of 0.5 in task arrival probability compared to existing methods such as DQN and DRL, respectively.

References

Almashhadani HA, Deng X, Abdul Latif SN, Ibrahim MM, Alshammari AH (2022) An edge-computing based task-unloading technique with privacy protection for Internet of connected vehicles. Wirel Pers Commun 127(2):1787–1808

Almashhadani HA, Deng X, Latif SNA, Ibrahim MM, AL-Hwaidi OHR (2023) Deploying an efficient and reliable scheduling for mobile edge computing for IoT applications. Mater Today Proc 80:2850–2857

Chi HR, Radwan A (2022) Fully-decentralized fairness-aware federated MEC small-cell peer-offloading for enterprise management networks. IEEE Trans Industr Inf 19(1):644–652

Dai X, Xiao Z, Jiang H, Lei M, Min G, Liu J, Dustdar S (2023) Offloading dependent tasks in edge computing with unknown system-side information. IEEE Trans Serv Comput 16(6):4345–4359

Fang C, Xu H, Yang Y, Hu Z, Tu S, Ota K, Yang Z, Dong M, Han Z, Yu FR, Liu Y (2022) Deep-reinforcement-learning-based resource allocation for content distribution in fog radio access networks. IEEE Internet Things J 9(18):16874–16883

Jalilvand Aghdam Bonab M, Shaghaghi Kandovan R (2022) QoS-aware resource allocation in mobile edge computing networks: using intelligent offloading and caching strategy. Peer-to-Peer Netw Appl 15(3):1328–1344

Liu Z, Liu Y, Lei Y, Zhou Z, Wang X (2023) ARLO: an asynchronous update reinforcement learning-based offloading algorithm for mobile edge computing. Peer-to-Peer Netw Appl 16(3):1468–1480

Mu S, Shen Y (2022) Stochastic learning for opportunistic peer-to-peer computation offloading in IoT edge computing. China Commun 19(7):239–256

Ren J, Hou T, Wang H, Tian H, Wei H, Zheng H, Zhang X (2024) Collaborative task offloading and resource scheduling framework for heterogeneous edge computing. Wirel Netw 30(5):3897–3909

Robles-Enciso A, Skarmeta AF (2023) A multi-layer guided reinforcement learning-based tasks offloading in edge computing. Comput Netw 220:109476

Tang Q, Liu L, Jin C, Wang J, Liao Z, Luo Y (2022) An UAV-assisted mobile edge computing offloading strategy for minimizing energy consumption. Comput Netw 207:108857

Tong SR, Yang CH (2022) Efficient broadcast scheduling at mobile cloud edges for supporting news-broadcast-on-demand over P2P streaming. Peer-to-Peer Netw Appl 15(3):1345–1356

Zeng F, Rou R, Deng Q, Wu J (2023) Parked vehicles crowdsourcing for task offloading in vehicular edge computing. Peer-to-Peer Netw Appl 16(4):1803–1818

Zhang X, Liu J, Zhang R, Huang Y, Tong J, Xin N, Liu L, Xiong Z (2023) Energy-efficient computation peer offloading in satellite edge computing networks. IEEE Trans Mob Comput 23(4):3077–3091

Zhou H, Wang Z, Zheng H, He S, Dong M (2023) Cost minimization-oriented computation offloading and service caching in mobile cloud-edge computing: an A3C-based approach. IEEE Trans Netw Sci Eng 10(3):1326–1338

Chapter 6
Energy-Efficient Resource Management and Clustering for the Wireless Networks Using Kent Mapping-Based Butterfly Optimization Algorithm

Sowmya Madhavan, G. S. Nijaguna, B. A. Smitha, R. Rana Veer Samara Sihman Bharattej, and R. Mohan Naik

Abstract Energy-efficient resource allocation for Fifth Generation (5G) and the wireless networks has become a primary research challenge because of enhancing the small cells (SC) concentrations and maximum Quality of Experience (QoE) necessities for the professional handlers. Make sure the QoE as well as energy effectiveness in important in mobile networks, however, these aims are continuously conflicting and hardly solved simultaneously in previous solutions. Hence, this research proposes the Kent mapping-based butterfly optimization algorithm (KM-BOA) approach for the energy-efficient resource management-based clustering for the wireless networks. The KM-BOA approach efficiently allocates resources such as bandwidth, power as well as computation to minimize an energy consumption in wireless networks, particularly for devices with constrained battery life. The proposed KM-BOA approach attains the better energy efficiency of 13.6, 13.9, 14.5, 14.9, and 15.1 based on the

---

S. Madhavan
Department of Electrical and Electronic Engineering, Nitte Meenakshi Institute of Technology, Nitte (Deemed to be University), Bengaluru, India
e-mail: sowmya.madhavan@nmit.ac.in

G. S. Nijaguna
Department of Information Science and Engineering, S.E.A. College of Engineering & Technology, Bengaluru, India
e-mail: nijagunags@seaedu.ac.in

B. A. Smitha
Department of Computer Science Engineering (Data Science), RNSIT Institute of Technology, Bengaluru, India

R. Rana Veer Samara Sihman Bharattej (✉)
Doctorate of Business Administration, National Louis University, Tampa, FL, USA
e-mail: rupavathrana@gmail.com

R. Mohan Naik
Department of Electronics and Communication Engineering, Shri Dharmasthala Manjunatheshwara Institute of Technology, Ujire, India
e-mail: mohannaik@sdmit.in

G. P. Ramesh et al. (eds.), *Empowering IoT with Distributed Edge Computing*, IoT and Wireless Communication Networks, https://doi.org/10.1007/978-981-96-9455-6_6

number of iterations of 2, 4, 6, 8, and 10, respectively, as compared to the existing methods like modified K-means clustering approach.

Keywords Butterfly optimization algorithm · Energy efficient · Kent mapping · Quality of experience · Resource allocation

Introduction

Through the developments in micro-electromechanical systems (MEMS) and wireless technology, the researchers were employed on the advancements named Internet of Things (IoT), big data analytics as well as cloud computing to address different real-life issues (Udayasankaran and Thangaraj 2023). Between all these advancements, IoT has developed as the most auspicious one because of its capability to associate IoT which allowing extraordinary computing capability. Since the inception of IoT, wireless sensor network (WSN) has always been an integral part of IoT (Chaurasiya et al. 2023). In this condition, devastating clustering and routing protocols are expected to increase the framework lifetime and attain an enhanced utilization of the power assets (Zeb et al. 2022). Moreover, the wired networks are most challenging and costly for designing as well as maintaining than the wireless networks. Thus, advanced networking mechanisms arranges the wireless communication due to these certainties. In this procedure, the number of sensors devoted large number of periods in a condition, but not performs effectively in communications (Firdous et al. 2022). Thus, in what manner to allocate the sensors states in a network to attain the effective data transmission is a challenging problem for target following areas (Venkatesan et al. 2022). To perform with this issue, the clustering approaches are broadly used as the means to enable supportive data processing as well as maintain the resources of the sensor networks (Mohajer et al. 2022). Various researchers have developed various clustering approaches for the various fields, between the most illustrative one is designing the on-demand clusters among the static clusters. An activated cluster is effectively created based on the location of the target moving as well as prediction mechanism is utilized to estimate the target trajectory (Ben Gouissem et al. 2022; Bal et al. 1242). The important highlights of this article are as trails:

- This research proposes a Kent mapping-based butterfly optimization algorithm (KM-BOA) approach for the energy-efficient resource management-based clustering for the wireless networks.
- The KM approach supports in making optimal clusters in wireless networks, make sure that resources are distributed efficiently between the nodes based on network topology as well as traffic constraint.
- The significance of the proposed KM-BOA approach for the resource allocation-based clustering is estimated by using the performance metric of energy efficiency based on number of iterations.

The supplementary portions of this manuscript are provided as tracks: Section "Literature Survey" provides the literature survey. Section "Proposed Methodology" provides a proposed methodology. Section "Experimental Results" describes the results and discussion. Section "Conclusion" depicts a conclusion.

Literature Survey

In this portion, the existing works related to the resource allocation and clustering are discussed, along with their advantages and limitations.

Ajay et al. (2022) implemented the clustering approach as well as collaborative resource allocation for machine-to-machine communication devices (MTCDs) resource management. The complexity of the resource allocation and the clustering was distinguished as the enhancement of the energy efficiency issue. The authors attained the effective power distribution plan by continuous energy efficiency maximization approach and then provided the modified K-mean clustering approach for solved the above-mentioned problems. Based on the significance of the non-linear fractional utilizations, the authors divided the problems into two categories such as redistribution of the power and cluster. However, a continuous nature of the energy efficiency maximization process resulted in delays, caused the responsiveness of the system.

Beshley et al. (2022) introduced the radio resource allocation as well as optimization approach to solve modifying user Quality of Experience (QoE) necessities and minimized the energy utilization in multi-layer 5G network. The introduced approach was varying from the familiae one such that it assumed the QoE necessities of business users and loas localization to effectively distributed a service procedure among the macro cells (MCs) and small cells (SCs). The introduced approach utilized the Voronoi architecture into energy-efficient design a 5G radio access network (RAN) through modifying the SCs to low-power mode while the users were not provided the dynamic operators. However, obtain an effective balance among meeting QoE for business users and minimized energy consumption was challenged because of conflicting objectives.

Ghafoor et al. (2022) maximized the energy efficiency through user equipment clustering (UE-C) through download hybrid non-orthogonal multiple access (H-NOMA) based Beyond 5G (B5G) HetNets. The authors formulated an optimization issue integrated an UE admission in cluster, UE integrated through the base station (BS) as well as power allocation supported through the H-NOMA. The issue formulated was a kind of non-linear concave fractional programming (CFP) issue. A Charnes-Cooper transformation (CCT) was used to an updated non-linear CFP issue for modify as the concave optimization. However, the procedure of transforming the CFP into a concave optimization issue utilized a CCT extended more overhead, minimized the scalability in large networks.

Bashir et al. (2023) presented the two new nature-encouraged approaches which considered the energy consumption of both CPU and memory during the virtual machine (VM) placement process. The presented approaches were related to the Artificial Bee Colony (ABC) as well as Particle Swarm Optimization (PSO) which does not utilize for scheduling the VM while the consideration of the energy consumption of CPU and memory. The service level agreement (SLA)-aware variants of the presented energy-efficient approaches were provided to the solve the problem of resultant SLA violations. However, the ABC and PSO approaches do not explicitly integrated the scheduling mechanisms for VMs.

Qu and Li (2022) introduced the Tracking Anchor-based Clustering Method (TACM), in that anchors were developed to contribute the stimulation suggestions for sensors based on a destination location. A Rough Fuzzy C-Means (RFCM) approach was utilized for trace anchor and utilized a membership chart for the stimulation of the sensors to create the clusters. Furthermore, the state of the cluster members (CM) was allocated through the utilization of 0–1 programming to minimize the duplicate transmissions. However, a significance of anchors in providing activation signs were effectively depended heavily on their precise placement.

Proposed Methodology

This research proposes the shared resource management architecture for the mobile-to-mobile (M2M) network. To maintain the resources of the systems, the recommended designs utilize the local resource controller (LRC) as well as global resource controllers (GRC) and the cooperative resource distribution and the clustering for MTCD, these are primary tasks of LRC as well as GRC.

Resource Management

Every LRC control involves BS or the individual MTCD. A GRC obtains the central data from the associated BS and the MTCD. The LRC obtains the resource allocation as well as clustering plan of the GRC. Figure 6.1 demonstrates the architecture of the cooperative resource management.

Administration of Global Resource. The input data are maintained through the GRC and this information is transmitted to the GRC through the LRC of the BS. This information involves the communication array, size of the network as well as greater number of CH and the CM. It collects the channel parameters of the MTCD's LRC, greater number of transmission power and the smaller number of transmission rate. Thus, the GRC understand the clustering and the power management of the MTCD.

Optimization Problem. An energy consumption of the MTCD is challenging because of the battery-powered sensors or the trivial devices through radio frequency identification (RFI). These batteries in the MTCDs are continuously challenging or

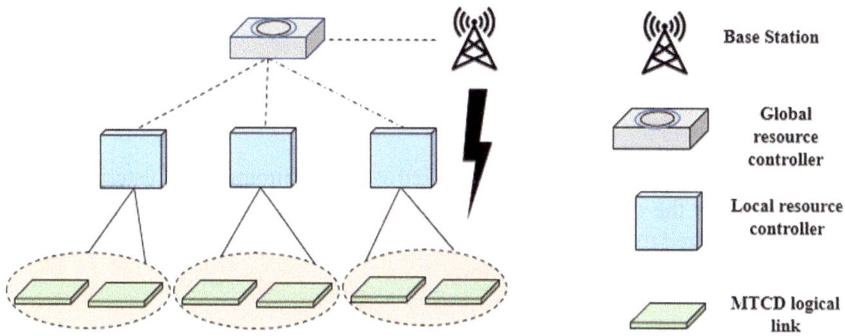

Fig. 6.1 Architecture of the cooperative resource management

unbearable to control. In that case, the operations are drop by the MTCD once one of the batteries are exhausts. Creating an energy effective data transmission system is challenging for minimum energy utilization and the long MTCD life time meanwhile the transmission of the data attains the large amount of energy. Promptly, the minimum power MTCD transmission effective should be make sure. Concentrating on an energy efficiency indicator enables of the trade-off among the transmission effectiveness and the energy utilization.

The development of all the MTCD in various data communication forms is extended each other. Through the examination of the transmission channel, rate of the data and the constraints of the transmission power of the MTCD, this task becomes an issue for an entire network.

The energy efficiency of the system is formulated in Eq. (6.1) as follows:

$$\Psi = \sum_{j=1}^{M} \Psi_i \tag{6.1}$$

where, Ψ_i denotes the energy consumption of the MTCD. The term for Ψ_i is formulated in Eq. (6.2) as follows:

$$\Psi_i = L_i^d \Psi_i^d + \sum_{l=1,l\neq i}^{M} \sum_{k=1}^{K_1} \alpha_{l,k} L_{i,k}^c \psi_{i,l}^c \tag{6.2}$$

where, $L_i^d \epsilon 0, 1$ denotes the MTCD_i transmitter as well as receiver's dynamic load balancing variable.

Limits on Optimization

This portion illustrates the limits on an effective utilization of the resources and the clustering form.

- The number of CH which whether utilized as the maximum: The clustering method must follow the maximum number of limitations of the CH. Whether the term $N_{\max}$ is utilized to refer to the CH, which postulates the maximum number of CH as formulated in Eq. (6.3) as follows:

$$K_1 \leq N_{\max} \rightarrow C1 \tag{6.3}$$

- Maximum number of clustered CM: Through considering that an individual CH is associated through the maximum number of CMs, which is formulated in Eq. (6.4) as follows:

$$\sum_{i=1}^{M} L_{i,k}^{c} \leq M_1,\ 1 \leq k \leq K_1 \rightarrow C2 \tag{6.4}$$

This research proposes the effective optimization approach to obtain the clustering. Details of this approach are provided in the following section.

Butterfly Optimization Algorithm

The BOA approach simulates the foraging behavior of butterflies to optimize energy-efficient resource management in wireless networks through balancing load distribution and minimizing energy consumption. It enhances the clustering efficiency through dynamically selecting optimal cluster heads and resource allocation strategies according to network conditions. Every butterfly in the BOA produces the particular scent. Simultaneously, the individual butterflies observe fragrances of the other (Alweshah et al. 2022; Alhasnawi et al. 2023). Based on the focus of the scent, every butterfly significantly reintroduces the location of butterfly group by two search approaches at search procedure. The butterfly's scent focus is formulated in Eq. (6.5) as follows:

$$f = cI^{a} \tag{6.5}$$

where, f denotes an intensity magnitude of a scent; c denotes a perceptual form; I denotes an incentive intensity; a denotes an energy exponent and ranges of a and $I = c$ lies in a range among 0 and 1, respectively. A search approach's selection is identified through the arbitrary parameter into the range among 0 and 1. An initial search approach utilizes the global search and these butterflies generates the scents at their position. Through comparing the scent of every butterfly, a butterfly by an effective scent is chosen and its location is observed as an effective location. In a global search plan, a butterfly group move over them through an effective scent for attain an aim of restorative a location of butterfly group. The particular search approach is formulated in Eq. (6.6) as follows:

$$x_i^{t+1} = x_i^t + \left(r^2 \times g^* - x_i^t\right) \times f_i \tag{6.6}$$

In Eq. (6.6), x_i^t denotes a space vector x_i for i th butterfly in number of iterations t. Here, g^* denotes a globally optimal individual in a present iteration. A scent of an i th butterfly is demonstrated through f_i and r denotes an arbitrary number in a range among 0 and 1, respectively. Another search approach is the local search. Assuming various conditions like distance among the butterfly individuals, an approach employs the local search is formulated in Eq. (6.7) as follows:

$$x_i^{t+1} = x_i^t + \left(r^2 \times x_j^t - x_k^t\right) \times f_i \tag{6.7}$$

In Eq. (6.7), x_i^t and x_k^t denote a solution vector of i th as well as k th butterfly in an optimal space and r denotes the arbitrary range between 0 and 1, respectively. Though the BOA involves the rapid search speed in a global search, when the butterfly individual through effective scent presents, various butterflies are located far rapidly. An approach solution procedure involves the worst range of optimal outcomes and is simply fall in local problems. A local search is based on the different arbitrary solution vectors; thus, an approach is simply missing the effective solution.

Kent Mapping. Chaos principle is considered through arbitrariness, non-repeatability, ergodicity as well as uniformity. As like to stochastic search, chaos principle employs the complete as well as entire search in an optimal solution. An overview of chaos principle into the metaheuristics approach significantly enhance the diversity of the population as well as enhance a solving effectiveness. A development of chaotic mapping in BOA enhances a convergence effectiveness of an approach, make sure the diversity of the butterfly locations, enhances the searching capability of an approach and solve local optimal problem in a procedure of an approach task. Integrating a mapped parameter r into a task effectively enhance the capability of BOA. Furthermore, the fixed mapping number is fixed to solve an entire deliberate iteration speed while enhancing a local optimal capability in a previous search. A particular expression is formulated in Eq. (6.8) as follows:

$$r = \begin{cases} (1-r)/(1-u) & r < u \\ \frac{r}{u} & r \geq u \end{cases} \tag{6.8}$$

A location regeneration procedure of the butterfly is optimized through the integrating the mapped r into a location regeneration estimation of butterfly group. The

BOA is improved through Kent mapping, converges quickly to the optimal solution, minimizing the time required for clustering and resource management.

Experimental Results

In this section, the effectiveness of the KM-BOA approach is implemented on Python 3.10.11 with the system configurations of intel i5 processor, 16 GB RAM and windows 10 OS. The simulation parameters of the proposed KM-BOA approach are outlined in Table 6.1.

Performance Analysis

The significance of the proposed KM-BOA approach is estimated and compared with the various existing approaches according to performance metric of energy efficiency. Table 6.2 demonstrates performance analysis of the energy efficiency in terms of number of iterations.

In Table 6.2, the performance analysis of an energy efficiency based on number of iterations for proposed KM-BOA approach. The number of iterations such as 2, 4, 6, 8 and 10 are considered for the estimation and validation purpose. The existing optimization approaches such as Particle Swarm Optimization (PSO), Cuckoo Search Optimization (CSO), Harris Hawks Optimization (HHO), and BOA are estimated and compared with the proposed KM-BOA approach. The proposed KM-BOA approach

Table 6.1 Simulation parameters of the proposed KM-BOA approach

Parameters	Values
Bandwidth	180 kHz
Transmission power	0.15 W
Number of MTCD	15
Energy consumption	0.3 W

Table 6.2 Performance analysis of energy efficiency (J) with respect to number of iterations

Methods	Number of iterations				
	2	4	6	8	10
PSO	10.2	10.6	12.1	14.6	15.3
CSO	10.3	10.4	11.4	14.2	14.4
HHO	11.3	12.3	13.1	13.5	13.9
BOA	12.3	12.5	12.9	13.1	13.6
KM-BOA	13.6	13.9	14.5	14.9	15.1

Table 6.3 Performance analysis of energy efficiency with respect to number of MTCD

Methods	Number of MTCD (bit/J)				
	15	30	45	50	75
PSO	5.92	6.03	6.65	6.76	6.87
CSO	4.28	4.56	3.56	3.89	5.10
HHO	3.28	3.67	3.12	3.48	4.78
BOA	2.38	2.66	2.98	3.22	4.46
KM-BOA	1.92	2.13	2.45	3.81	4.13

attains the better energy efficiency of 13.6 J, 13.9 J, 14.5 J, 14.9 J, and 15.1 J based on the number of iterations of 2, 4, 6, 8, and 10, respectively.

Table 6.3 demonstrates the performance analysis of an energy efficiency based on number of MTCD. A significance of the proposed KM-BOA approach is estimated and compared with the various existing approaches like PSO, CSO, HHO, and BOA, respectively. As compared to these existing approaches, the proposed KM-BOA approach attains the better efficiency of 1.92 bit/J, 2.13 bit/J, 2.45 bit/J, 3.81 bit/J, and 4.13 bit/J based on the number of MTCD, respectively. The KM approach supports in designing optimal clusters in wireless networks, ensuring that resources are distributed effectively between the nodes based on network topology as well as traffic constraint.

Discussion

This section discusses the limitations of the existing works and how the proposed method tackles these problems, along with their advantages. The limitations of the existing works are: In (Bal et al. 1242), a continuous nature of the energy efficiency maximization process resulted in delays, caused the responsiveness of the system. In (Ajay et al. 2022), obtain an effective balance among meeting QoE for business users and minimized energy consumption was challenged because of conflicting objectives. In (Beshley et al. 2022), the procedure of transforming the CFP into a concave optimization problem utilized the CCT extended more overhead, minimized the scalability in large networks. the ABC and PSO approaches (Ghafoor et al. 2022) do not explicitly integrated the scheduling mechanisms for VMs. In RFCM approach (Bashir et al. 2023), a significance of anchors in providing activation signs were effectively depended heavily on their precise placement. Hence, this research proposes the KM-BOA approach for the effective energy-efficient resource allocation-based clustering in wireless networks. The KM approach supports in designing optimal clusters in wireless networks, make sure that resources are distributed effectively among nodes based on network topology as well as traffic constraint. The proposed KM-BOA approach attains the better energy efficiency of 13.6, 13.9, 14.5, 14.9, and 15.1 based on the number of iterations of 2, 4, 6, 8, and 10, respectively. Furthermore, the proposed KM-BOA approach attains the better efficiency of 1.92 bit/J, 2.13 bit/J, 2.45 bit/J, 3.81 bit/J, and 4.13 bit/J based on the number of MTCD, respectively.

Conclusion

This research discovers the effective resource allocation and clustering in the wireless networks. This research proposes the KM-BOA approach for the energy-efficient resource allocation-based clustering in wireless networks. Moreover, this research illustrates the collaborative strategic plan architecture, pursued through the strategy for enhancing the system effectiveness through the cooperative resource allocation and the clustering. KM-BOA dynamically familiarizes to changes in network circumstances, such as different traffic load, node mobility or interference through re-optimizing resource allocation and cluster formation. The BOA is improved through Kent mapping, converges quickly to the optimal solution, minimizing the time required for clustering and resource management. This flexibility ensures that the network controls an optimal performance as well as energy efficiency even in dynamic environments with changing network demands. The proposed KM-BOA approach attains the better energy efficiency of 13.6, 13.9, 14.5, 14.9, and 15.1 based on the number of iterations of 2, 4, 6, 8, and 10, respectively. The future work will involve the hybrid optimization algorithm to enhance the overall system performance.

References

Ajay P, Nagaraj B, Jaya J (2022) Algorithm for energy resource allocation and sensor-based clustering in M2M communication systems. Wirel Commun Mob Comput 2022(1):7815916

Alhasnawi BN, Jasim BH, Bureš V, Sedhom BE, Alhasnawi AN, Abbassi R, Alsemawai MRM, Siano P, Guerrero JM (2023) A novel economic dispatch in the stand-alone system using improved butterfly optimization algorithm. Energ Strat Rev 49:101135

Alweshah M, Khalaileh SA, Gupta BB, Almomani A, Hammouri AI, Al-Betar MA (2022) The monarch butterfly optimization algorithm for solving feature selection problems. Neural Comput Appl 34:11267–11281

Bal PK, Mohapatra SK, Das TK, Srinivasan K, Hu YCL (2022) A joint resource allocation, security with efficient task scheduling in cloud computing using hybrid machine learning techniques. Sensors 22(3):1242

Bashir S, Mustafa S, Ahmad RW, Shuja J, Maqsood T, Alourani A (2023) Multi-factor nature inspired SLA-aware energy efficient resource management for cloud environments. Clust Comput 26(2):1643–1658

Ben Gouissem B, Gantassi R, Hasnaoui S (2022) Energy efficient grid based k-means clustering algorithm for large scale wireless sensor networks. Int J Commun Syst 35(14):e5255

Beshley M, Kryvinska N, Beshley H (2022) Energy-efficient QoE-driven radio resource management method for 5G and beyond networks. IEEE Access 10:131691–131710

Chaurasiya SK, Mondal S, Biswas A, Nayyar A, Shah MA, Banerjee R (2023) An energy-efficient hybrid clustering technique (EEHCT) for IoT-based multilevel heterogeneous wireless sensor networks. IEEE Access 11:25941–25958

Firdous S, Bibi N, Wahid M, Alhazmi S (2022) Efficient clustering-based routing for energy management in wireless sensor network-assisted internet of things. Electronics 11(23):3922
Ghafoor U, Khan HZ, Ali M, Siddiqui AM, Naeem M, Rashid I (2022) Energy efficient resource allocation for H-NOMA assisted B5G HetNets. IEEE Access 10:91699–91711
Mohajer A, Sorouri F, Mirzaei A, Ziaeddini A, Rad KJ, Bavaghar M (2022) Energy-aware hierarchical resource management and backhaul traffic optimization in heterogeneous cellular networks. IEEE Syst J 16(4):5188–5199
Qu Z, Li B (2022) An energy-efficient clustering method for target tracking based on tracking anchors in wireless sensor networks. Sensors 22(15):5675
Udayasankaran P, Thangaraj SJJ (2023) Energy efficient resource utilization and load balancing in virtual machines using prediction algorithms. Int J Cogn Comput Eng 4:127–134
Venkatesan VK, Izonin I, Periyasamy J, Indirajithu A, Batyuk A, Ramakrishna MT (2022) Incorporation of energy efficient computational strategies for clustering and routing in heterogeneous networks of smart city. Energies 15(20):7524
Zeb A, Wakeel S, Rahman T, Khan I, Uddin MI, Niazi B (2022) Energy-efficient cluster formation in IoT-enabled wireless body area network. Comput Intell Neurosci 2022(1):2558590

Chapter 7
Logistic Map-Based Gazelle Optimization Algorithm for Energy-Aware Mobile Edge Computing

Vishwanath Petli, T. Anuradha, J. Sujatha, N. Geetha, and S. Shailaja

Abstract Mobile edge computing (MEC) has currently recognized as capable computing model to provide a delay-sensitive, computation-intensive services for mobile users. The existing methods suffered from high delay and energy consumption (EC). Therefore, the logistic map-based gazelle optimization algorithm (LMGOA) is proposed to reduce EC by optimizing the use of a task offloading solution and resource management in a highly dynamic environment. The logistic map used for initialization is incorporated to reinforce the disorder and equality of the initial solution across the space, cause a superior optimization in search space. This helps in preventing the algorithm from converging earlier to a solution and enhance the algorithms capability to find the best solutions for energy-efficient MEC. The delay, EC and throughput are taken to estimate the LMGOA performance. The LMGOA achieved delay of 32.65 s, EC of 16.509 J and throughput of 7692.3 Mbps for 1000 no. of mobile devices which is better than existing approaches.

Keywords Energy consumption · Gazelle optimization algorithm · Logistic map · Mobile devices · Mobile edge computing

V. Petli
Department of Electronics and Communication Engineering, H.K.E. Society's Sir M. Visvesvaraya College of Engineering, Raichur, India

T. Anuradha (✉) · S. Shailaja
Department of Computer Science and Engineering, PDA College of Engineering, Kalaburagi, Karnataka, India
e-mail: anuradhat@pdaengg.com

S. Shailaja
e-mail: shailajas@pdaengg.com

J. Sujatha
Sir M. Visvesvaraya College of Engineering Raichur, Affiliated to VTU, Belagavi, India

N. Geetha
Department of Computer Science and Engineering, Sir M. Visvesvaraya College of Engineering Raichur, Affiliated to VTU, Belagavi, India

G. P. Ramesh et al. (eds.), *Empowering IoT with Distributed Edge Computing*, IoT and Wireless Communication Networks, https://doi.org/10.1007/978-981-96-9455-6_7

Introduction

The unmanned aerial vehicles (UAVs) have recently attracted numerous attention due to their uses and services such as augmented and virtual reality, smart manufacturing inspection Internet of Vehicles (IoV), video and 3D gaming, and e-health (Alharbi et al. 2023). This is inexpensive and has significant implementation latency because of data transmission and Cloud Computing (CC) and is not suitable for real-time applications (Masdari et al. 2022). To overcome this problem, there is an up-and-coming approach known as MEC which offloads computation and storage resources closer to mobile device of the mobile networks allowing the performance of complex applications on the mobile device while meeting real-time constraints (Pérez et al. 2021). MEC employs edge data centers (EDCs) which are data centers located close to data sources thus reduce dependence on conventional centralized data centers hence reducing traffic congestion and delay (Huang et al. 2023; Xu et al. 2020). The MEC enhances energy efficiency as it enables users to transfer complicated calculations to edge server, which in contrast minimizes energy consumption (EC) of user devices and coping with overcrowded networks which are characteristic for CC (Baker et al. 2024). The communication, computation, storage resource, and management are necessary to provide cost-effective and dependable edge network services (Lee et al. 2024). The edge offers a distributed network solution where various admin domains join a permissioned network and are represented with only one node which is a major challenge to edge computing (EC) security (Mehrabi et al. 2021). An accurate MEC model facilitates resource providers to estimate the energy usage thereby enabling effective scheduling in MEC framework (Hossam et al. 2024). The contributions are stated as follows:

- The logistic map-based gazelle optimization algorithm (LMGOA) is proposed to minimize EC by optimizing the task offloading and resource management in highly dynamic environments.
- The logistic map is introduced into the initialization phase to enhance the diversity and uniformity of the initial solution across search space thereby enabling better exploration and optimization.
- The LMGOA helps to prevent premature convergence to a solution space and enhance the algorithms capability to identify optimal solutions for energy-efficient MEC.

The remaining stages are formed as follows: Section "Literature Review" demonstrations literature review, Section "Proposed Method" shows proposed method, Section "Result Analysis" depicts results and discussion, and Section "Conclusion" signifies conclusion.

Literature Review

Li et al. (2023) developed an Energy-Aware and Trust-Collaboration Cross-Domain Resource Allocation (ETCRA) to edge-cloud environments. The aim was for reduce overall value of the CSF and at the same time optimize latency in flow of work and confidence of cross edges. The dynamic approaches were suggested for arrive at desirable decision regarding the task-resource mapping. It consisted of two phases such as the quantitative decision-making of primary resource provision decision-making using Particle Swarm Optimization (PSO) statically and dynamic apprising decision-making of trust value valuation.

Li et al. (2018) suggested an Energy-aware task offloading mechanism in multiuser mobile edge CC. The offloading decision was defined as 0–1 nonlinear integer programming issue constrained with channel interference limit and time limit. Based on organization and importance resolve of devices, opposite auction-based offloading was developed which aimed at the solution of optimization issues to enhance energy proficiency. The developed model not only gives offloading decision, then it similarly helps in resource allocation. Analyzing energy efficiency performance, the simulation discovered benefit of developed scheme over other methods.

Avgeris et al. (2022) introduced a distributed energy-aware resource allocation at Edge named as ENERDGE. The trade-off between low total EC, end-to-end delay constraints, load balancing at the Edge a concept called Markov random field-based mechanism for workload distribution was introduced. The proposed approach analyzed the practical case that considers the diverse types of mobile applications, heterogeneous edge devices with different computational power, user mobility and their behavior affecting the nature of wireless link. The framework was subsequently augmented with a prediction mechanism that balanced they managed the physical resources logistics.

Li et al. (2021) developed an Energy-aware task offloading with deadline constraint (DRL-E2D) in mobile-EC. The DRL-E2D was developed to maximize the reward under task's limit constraints. Within actor-critic approach, action was integrated to DRL-E2D for address higher separate action-space issues. A less-complexity of K-nearest neighbor (KNN) was applied as approximation for extract better discrete-actions from constant action space. The output demonstrated DRL-E2D outperformed comparison algorithms across all parameter settings which leads to state changes in MEC environment.

Zhou et al. (2021) developed an Edge Intelligent Energy Efficient Model in MEC named as ECMS. The ECMS adopted Elman Neural Network (ENN) and feature selection for accurate modeling of MEC on edge servers. ECMS evaluated 29 factors associated with edge server MEC and used ENN to build model. The ECMS was able to manage load fluctuations and different types of tasks like CPU-bound tasks, online transactions and I/O-bound tasks. The ECMS estimating its accuracy and time required to train the model and comparing it with several baseline strategies.

Proposed Method

The mobile edge computing (MEC) system depicted in Fig. 7.1 which is a mobile-server-assisted MEC system. The system comprises the following things:

- Coordination Center: It tracks the current state of system management, collecting user queries and making appropriate choices relating to the delegation of tasks.
- Fixed Edge Servers: It is represented by the set notation $N = \{1, 2, \ldots, N\}$.
- Mobile Servers: It is noted as $M = \{1, 2, \ldots, M\}$.
- Mobile Users: It is defined as a set $K = \{1, 2, \ldots, K\}$.

Each edge server is positioned through BS and linked to other by large bandwidth wired links. As the rate of transmission delay between them is negligible, we integrate each base station and its related server, characterized as $n \in N$. Every BS operates a particular cellular area (cell) which does not share the same cell with other BS. A fixed edge server is a computation service provider for the mobility user and these servers are connected with each other through high bandwidth optical fibers forming a Metabolite Area Network (MAN). These fixed servers are in the shape of a complete connected, single-hop network that perform high-speed data interchange. The Mobil serving nodes are intelligent wireless devices or vehicles that have spare computational capacity and are in field of base stations. They link through cellular and help in serving the mobile users hence all mobile and fixed servers are defined through the $S = N \cup M = \{1, 2, \ldots, S\}$ where S denotes total number of servers. The computing capability of a server $s \in S$ is represented by f_s and it is considered as heterogeneous.

It works in a cycle-by-cycle basis. In this case, each cycle is represented as $t \in T = \{1, \ldots, T\}$. At the onset of every cycle, the management platform is assumed

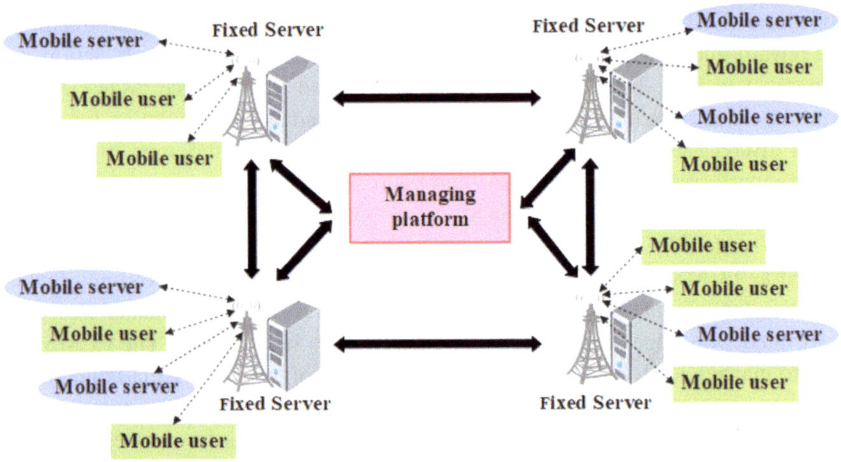

Fig. 7.1 MEC architecture

to have prior knowledge of all tasks that are likely to come during a given cycle. We note let R^t denote the set of tasks arriving in cycle t. It is noted that any user $k \in K$ create multiple tasks in a cycle. To simplify the approach, multiple tasks assigned to one user are treated as a virtual separation of a user and, therefore, one-to-one correspondence between the two sets is created. In case of a specific task $r_k \in R^t$, associated with the user $k \in K$, the task set is defined by $R^t = \{r_1, \ldots, r_k\}$. Every user has its POIs and perform movements between them randomly while staying at any of them for some time. The collective of the mobile servers associated with a fixed server $n \in N$ under the base station coverage in cycle t is represented as $M_n^t \subseteq M$. The mobile servers are portable and many of them are run on batteries with differing capacities to suffice their most basic specifications. If the residual energy level within a mobile server is below a certain predetermined limit, the server ends to help with computations. The management platform coordinates assignments and distributes corresponding tasks to servers depending on detailed information about the assigned task, energy of the servers and condition of the network. In other words, the MEC is managed in a dynamic way so that mobile servers sustained for as long as possible while avoiding long overall delay.

Delay Model

In edge network, every task $r_k \in R^t$ during time cycle t is offloaded using next four modes:

- Local Fixed Edge Server: The outcome is that the task is directly offloaded to local fixed edge server $n \in N$ to computation.
- Local Mobile Server: The task is then outsourced to a mobile server $m \in M_n^t$ within the same cell. This has the capability of transferring through the corresponding BS n with two hops across the wireless cellular system.
- Remote Fixed Edge Server: The task is delegated to one of the fixed edge servers $n' \in N\{n\}$. This needs to transmitted wirelessly through the affiliated BS n one hop to n, after which through wired connection to remote constant edge server.
- Remote Mobile Server: It is then outsourced to a remote mobile server $m \in M_n^t$. The flow path is through the local BS n, the remote BS n and through wired/ wireless channels and a remote mobile server m.

For these four offloading modes, the transmission delay experienced through a task is calculated as follows:

For first scenario, it is equivalent to one-hop delay from user to BS. For second case, the transmission delay was equal to a two-hop route is associated with BS. Then, it was equal to wireless link transmission delay of sending a message from the user to a BS corresponding to user and wired link transmission delay of sending the message from BS to a BS that contains remote fixed server. In quarter instance, it was equal to transmission delays of dual wireless links and one wired link between nodes in wireless line. Because all fixed servers were interconnected using high-speed wired link are elaborated with network.

Energy Model

The battery-powered mobile servers have to schedule personal energy level reserved for their operation limited by capacity of batteries involved. A mobile user assisted computing service, if the residual energy was prearranged and reserved level. However, the fixed servers typically getting power from the power grid and has the capability to offer uninterrupted computed services. Therefore, equality in considerations of the energy among the mobile servers for the task became a sensitive issue to the reinforcement of their operational time. This approach attempted to make the time of system service delays maximize the total service capability of the entire MEC system. For a mobile server *m*, the $E_m^{\max}$ is used for representing the initial energy amount while E_m^{self} is used for representing the personalized reserved energy.

Proposed LMGOA

The new optimization algorithm namely GOA is developed based on the behavior of gazelles such as grazing when there is no predator and fleeing incidentally when there is a predator is sighted. The proposed new GOA, called the IGOA, outperforms the original GOA by increasing global search accuracy and the convergence rate achieved with the new logistic mapping initialization (Wu et al. 2024). The LMGOA is used for energy-aware MEC through optimizing resource allocation and task offloading. The logistic map is applied in the initialization stage to ensure distributed and diverse population thereby enhancing the exploration of search space. This ensures the recognition of energy-efficient solutions, minimizing EC while maintaining optimal performance in MEC. Stimulated through survival behavior of gazelles avoiding hunters during foraging, optimization procedure of GOA is categorized into two phases: foraging for food when no predator, predator is seen or heard (exploitation stage) and fleeing when have seen or heard a predator, predator (exploration stage).

Logistic Map Initialization. To enhance randomness and uniformity of initial solution set in the process of GOA optimization algorithm, this research incorporates logistic mapping. The logistic map is expressed as Eq. (7.1),

$$x_{i+1} = x_i + \varphi \times (1 - x_i) \tag{7.1}$$

where, φ is a logistic coefficient.

Exploitation Stage. The gazelles graze with no predator in the vicinity and during this activity they are believed to move in a Brownian fashion. The position update method for an individual gazelle is defined as Eq. (7.2),

$$x_i^{p1} = x_i + v \cdot r_1 \cdot R_b(\text{Elite} - R_b \times x_i) \tag{7.2}$$

where x_i^{p1} and x_i are coordinates of ith gazelle earlier and next first location inform stage, the position vector of Brownian motion as Eq. (7.3),

$$f_b(x, \mu, \sigma) = \frac{1}{\sqrt{2\pi\sigma}} \exp\left(-\frac{(x - \mu^2)}{2\sigma^2}\right) = \frac{1}{\sqrt{2\pi}} \exp\left(-\frac{x^2}{2}\right) \tag{7.3}$$

where, Gaussian probability distribution function of position vector of Brownian motion is indicated by $f_b(x, \mu, \sigma)$ having $\mu = 0$ and $\sigma^2 = 1$.

Exploration Stage. If one recognizes a hunter, the gazelle performs circles with its tail and passes on them, while the hunter chases after it. Since both movements are contain direction mutations, a directional characteristic variable is qualitative and is present in both movements. In any cycle, no of iteration is odd gazelle moves in one direction and no of iteration is even the gazelle moves in the second direction. Specifically, some of the gazelles who spotted the hunter first initiate the response and the position updates of which conform to the Lévy flight distribution. The position update formula for gazelle during this phase is as Eqs. (7.4) and (7.5),

$$x_i^{p2.1} = x_i + v \cdot \mu_g \cdot r_2 \cdot R_l(\text{Elite} - R_l \times x_i) \tag{7.4}$$

$$x_i^{p2.2} = x_i + v \cdot \mu_g \cdot c_f \cdot R_b(\text{Elite} - R_l \times x_i) \tag{7.5}$$

Here, $x_i^{p2.1,2}$ signifies position in jth dimension of ith gazelle after next position update phase and c_f represents cumulative effect of hunter, which is calculated as Eq. (7.6),

$$c_f = \left(1 - \frac{m}{M}\right)^{2m/M} \tag{7.6}$$

where, m is a current number of iterations, M is maximum no. of iterations. The ability of gazelle escape from hunter's pursuit is described by Eqs. (7.7) and (7.8),

$$x_i^{p2.3} = \begin{cases} x_i + c_f[l_b + r_3 \cdot (u_b - l_b)] \cdot Q & \text{if } r' \le \text{psrs} \\ x_i + [\text{psrs}(1 - r_4) + r_4](x_{r_1} - x_{r_2}) & \text{else} \end{cases} \tag{7.7}$$

$$U = \begin{cases} 0, \text{ if } r_4 < 0.34 \\ 1, \text{ else} \end{cases} \tag{7.8}$$

Here, u_b and l_b denote higher and minor position limits of gazelle, respectively;r_3 and r_4 are the random numbers, psrs is escape rate and Q is a binary vector.

Result Analysis

The proposed LMGOA is simulated in Python with software requirements of intel i7 processor, windows 10 OS and RAM 8 GB. The delay, EC, and throughput are taken to estimate the LMGOA performance. Table 7.1 shows the LMGOA performance in terms of delay with various no. of mobile devices like 50, 200, 500, 800, and 1000. The Osprey Optimization Algorithm (OOA), Whale Optimization Algorithm (WOA) and GOA are analyzed and compared with LMGOA. The LMGOA achieved delay of 20.54, 22.71, 25.80, 29.46, and 32.65 s for 50, 200, 500, 800, and 1000 no. of mobile devices.

Table 7.2 shows the LMGOA performance in terms of EC with various no. of mobile devices like 50, 200, 500, 800, and 1000. The OOA, WOA, and GOA are analyzed and compared with LMGOA. The LMGOA achieved EC of 4.257, 6.481, 9.826, 12.457, and 16.509 J for 50, 200, 500, 800, and 1000 no. of mobile devices. However, as number of mobile device users increase rapidly and the traffic for which its performance reduces. In particular, the proposed method achieves slightly lower EC than the other three approaches.

Table 7.3 shows the LMGOA performance in terms of throughput with various no. of mobile devices like 50, 200, 500, 800, and 1000. The OOA, WOA, and GOA are analyzed and compared with LMGOA. The LMGOA achieved throughput of

Table 7.1 LMGOA performance in terms of delay (s)

No. of mobile devices	OOA	WOA	GOA	LMGOA
50	31.67	27.14	23.49	20.54
200	35.54	30.61	26.18	22.71
500	38.07	34.92	30.51	25.80
800	42.59	37.06	34.20	29.46
1000	47.62	41.57	37.09	32.65

Table 7.2 LMGOA performance in terms of EC (J)

No. of mobile devices	OOA	WOA	GOA	LMGOA
50	15.369	11.256	7.368	4.257
200	17.258	14.359	9.256	6.481
500	19.147	17.465	12.405	9.826
800	22.456	20.578	15.753	12.457
1000	27.159	24.954	19.672	16.509

Table 7.3 LMGOA performance in terms of throughout (Mbps)

No. of mobile devices	OOA	WOA	GOA	LMGOA
50	4726.4	5139.4	5436.2	6127.6
200	4938.5	5529.6	5864.7	6529.1
500	5283.9	5829.1	6135.4	6859.8
800	5619.7	6237.5	6459.8	7243.2
1000	5964.2	6626.6	6915.5	7692.3

6127.6, 6529.1, 6859.8, 7243.2, and 7692.3 Mbps for 50, 200, 500, 800, and 1000 no. of mobile. The rates of mobile are also determined by comparison of this proposed method, together with the other three methods. In a failure of processing that not specify information uploading, throughput continually retains a zero value. As the numeral of mobile devices is increasing, throughput received using proposed method is constantly higher than another approaches.

Discussion

Through mimicking the hunting and escape behavior from predators of gazelles, the LMGOA proposed to maintain optimal level of exploration and exploitation. The different linguistic starting points provided by the logistic map initialization enhances the search phase as a wide number of initial candidate solutions is generated which enhances the flexibility of the algorithm. The LMGOA mimicking the gazelle's foraging and escaping characteristics which makes it possible to force the algorithm to exploit promising areas of the space while at the same time diversifying once in a while to avoid premature convergence. It helps to prevent overloading servers that is low in energy which helping to distribute energy across the MEC system. Because it increases the global searches, save energy and prevent network overload, which is considered as an important for improving MEC systems. As a result, the gazelle-based behavior and the new initializing scheme of logistic map, the performance and energy efficiency of the LMGOA exceed the current state of the art for MEC systems.

Conclusion

The LMGOA is proposed to reduce complexity by optimizing the usage of a task offloading solution and resource management in high dynamic environment. The logistic map used for initialization is combined to reinforce the disorder and equality of the primary solution across the space, that leads to superior optimization in search space. This helps to prevent from early convergence to a solution and enhance the capability to find the optimal solutions for energy-efficient MEC. The LMGOA mimicking the gazelle's foraging and escaping characteristics which makes it possible to force algorithm to exploit promising areas of the space while at the same time diversifying once in a while to avoid premature convergence. The delay, EC, and throughput are taken to estimate the LMGOA performance. The LMGOA achieved delay of 32.65 s, EC of 16.509 J and throughput of 7692.3 Mbps for 1000 no. of mobile devices which is better than existing approaches. In future, hybrid meta-heuristic algorithm will be applied to further enhance the model performance.

References

Alharbi HA, Aldossary M, Almutairi J, Elgendy IA (2023) Energy-aware and secure task offloading for multi-tier edge-cloud computing systems. Sensors 23(6):3254

Avgeris M, Spatharakis D, Dechouniotis D, Leivadeas A, Karyotis V, Papavassiliou S (2022) ENERDGE: distributed energy-aware resource allocation at the edge. Sensors 22(2):660

Baker T, Al Aghbari Z, Khedr AM, Ahmed N, Girija S (2024) EDITORS: energy-aware dynamic task offloading using deep reinforcement transfer learning in SDN-enabled edge nodes. Internet Things 25:101118

Hossam HS, Abdel-Galil H, Belal M (2024) An energy-aware module placement strategy in fog-based healthcare monitoring systems. Clust Comput 27:7351–7372

Huang X, Lei B, Ji G, Zhang B (2023) Energy criticality avoidance-based delay minimization ant colony algorithm for task assignment in mobile-server-assisted mobile edge computing. Sensors 23(13):6041

Lee J, Kim Y, Keum D, Lee GM, Kim S, Han MH (2024) A design and implementation of energy-aware resilience architecture for mobile edge cloud. Environments 4:5

Li L, Zhang X, Liu K, Jiang F, Peng J (2018) An energy-aware task offloading mechanism in multiuser mobile-edge cloud computing. Mob Inf Syst 2018(1):7646705

Li Z, Chang V, Ge J, Pan L, Hu H, Huang B (2021) Energy-aware task offloading with deadline constraint in mobile edge computing. EURASIP J Wirel Commun Netw 2021:56

Li J, Qin Z, Liu W, Yu X (2023) Energy-aware and trust-collaboration cross-domain resource allocation algorithm for edge-cloud workflows. IEEE Internet Things J 11(4):7249–7264

Masdari M, Majidzadeh K, Doustsadigh E, Babazadeh A, Asemi R (2022) Energy-aware computation offloading in mobile edge computing using quantum-based arithmetic optimization algorithm, pp 1–26

Mehrabi M, Shen S, Hai Y, Latzko V, Koudouridis GP, Gelabert X, Reisslein M, Fitzek FH (2021) Mobility-and energy-aware cooperative edge offloading for dependent computation tasks. Network 1(2):191–214

Pérez S, Arroba P, Moya JM (2021) Energy-conscious optimization of edge computing through deep reinforcement learning and two-phase immersion cooling. Futur Gener Comput Syst 125:891–907

Wu D, Wu L, Wen T, Li L (2024) Microgrid operation optimization method considering power-to-gas equipment: an improved gazelle optimization algorithm. Symmetry 16(1):83

Xu C, Zheng G, Tang L (2020) Energy-aware user association for NOMA-based mobile edge computing using matching-coalition game. IEEE Access 8:61943–61955

Zhou Z, Shojafar M, Abawajy J, Yin H, Lu H (2021) ECMS: an edge intelligent energy efficient model in mobile edge computing. IEEE Trans Green Commun Netw 6(1):238–247

Chapter 8
An Energy-Aware Computation Offloading Task Using Meerkat Clan Algorithm with Chaotic Map and Crossover Strategy for Edge Computing

M. G. Kavitha, H. A. Vidya, and K. Anusha

Abstract In edge computing, the task offloading balance effectively, adaptability and reliability which aims to enhance the resource utilization, enhance the user experience while considering dynamic and limited resource. The Meerkat Clan Algorithm with Chaotic Map and Crossover Strategy (MCA-CC) is proposed in this research for energy-aware computation offloading in edge computing. The MCA-CC improves the exploration ability of the algorithm by adding chaotic maps, thus reducing the possibility of early convergence to a local solution. Chaotic maps bring in non-linear randomness into the search space and this enhances the utilization of several solution spaces efficiently. This feature is most useful when the search space is multimodal or has a non-convex shape when global optimum is difficult to locate. The crossover strategy helps in creating better inherited solutions and makes a switch to better solutions faster in the given problem space. Also, crossover is applied with chaotic map making it possible to enhance the level of exploration without compromising on the exploitation and to avoid trapping in exploitation suboptimal areas. The MCA-CC obtains execution time of 530 s, energy consumption (EC) of 85 mJ and service cost of 195$ which is better than state-of-art approaches.

Keywords Chaotic map · Computation offloading · Crossover strategy · Edge computing · Meerkat Clan algorithm

M. G. Kavitha · K. Anusha (✉)
Department of Computer Science and Engineering, Kalpataru Institute of Technology, Tiptur, India
e-mail: anushakit27@gmail.com

M. G. Kavitha
e-mail: kavitha.mg@kittiptur.ac.in

H. A. Vidya
Department of Computer Science and Engineering, Kalpataru Institute of Technology, Visvesvaraya Technological University, Belagavi, India

G. P. Ramesh et al. (eds.), *Empowering IoT with Distributed Edge Computing*, IoT and Wireless Communication Networks, https://doi.org/10.1007/978-981-96-9455-6_8

Introduction

The concept of multi-access edge computing (MEC) indicates placing computational and storage resources near the mobile end devices, preferably at the cellular base stations (BSs) and WiFi access points (Suganya et al. 2024). Accordingly, MEC is in line with the general trend of overlaying edge networks with communication, computing, and storage services (Mehrabi et al. 2021). To provide these services without intervention and at a low cost, it is therefore necessary to subdivide edge network resources communication, computation, storage, and organize them efficiently (Silva et al. 2021). Although cloud server execution has certain benefits like others, they are really challenged by higher latency problems and securities issues (He et al. 2020). In order to overcome this problem, edge computing is a concept where computation and storage capabilities are brought nearer to the network edges thereby bringing the perception of latency down (Alharbi et al. 2023). This has given rise to computation offloading and this makes it possible for intensive or delay sensitive computations to offloaded and performed in other remotely powerful devices (Sellami et al. 2022a). Moreover, the deployment of MEC applications requires much higher service capacity over networking, thereby creating capacity to offer multi-dimensional options characterizing MEC systems (Almuseelem 2023). While MEC minimizes energy consumption from computation on IoT devices, offloading computation intensive or energy hungry operations to MEC sever incurs extra energy cost for data transmission (Sellami et al. 2022b).

Zhao et al. (2021) suggested an iterative algorithm is presented to tackle them step by step. In particular, closed-form expressions for the offloading ratios, utilize the equivalent parametric convex optimization technique for calculating the power control policy and by applying the primal–dual algorithm for the efficient allocation of subcarrier and computing resources. Baker et al. (2024) developed an energy-aware dynamic task offloading utilizing deep reinforcement transfer learning (DRTL) in software-defined network (SDN) named as EDITORS for edge computing environments. In EDITORS, a task offloading framework will be designed with incorporating trusted edge nodes in mind, following the important features like energy efficiency, low latency, reliability, and flexibility. The DRTL agents are located at edge nodes where they work with the SDN controller to periodically update those offloading decisions based on the current conditions within the network as well as available resources.

Zhang et al. (2022) presented a real-time energy-aware offloading scheme for MEC. Since the problem of computation offloading and resource allocation is modeled as Mixed-Integer Nonlinear Problem (MINLP), this article employs a bi-level optimization to decouple the MINLP into two simpler problems. In addition, according to the flexible mobility of vehicle users (V-UEs) and the dynamism in the cloud computing platform, an offloading scheme using deep reinforcement learning (DRL) is employed to assist users to initiate the optimal offloading actions. Sada et al. (2024) implemented a LITOSS which was two-stage framework for task allocation at the scheduler level through a genetic algorithm (GA) and for edge server selection through a reinforcement learning (RL) agent. The experiments with Raspberry Pi and multiple edge servers to check the performance of Liverpool Airport framework show that framework provides better average accuracy. Li et al. (2021) introduced an energy-aware task offloading with deadline constraint (DRL-E2D) for multi-eNB MEC and based on DRL. To achieve the maximum reward when the task needs to be done within a certain amount of time. Finally, simple k-nearest neighbor method is used as an approximate function to obtain the best discrete action from a continuous action domain. The contributions are given as follows:

- The MCA-CC integrates the chaotic map to introduce non-linear randomness, improves the exploration ability and reducing the premature convergence in offloading and edge computing.
- By using chaotic maps, the algorithm effectively transverse various solution space which ensures effective search process.
- The crossover strategy enhances the exploration capabilities and helps to share information among candidate solutions which leads to optimize task offloading and resource allocation.

The rest portions are explained in the following way: Section "Proposed Methodology" explains the proposed methodology, Section "Result Analysis" elaborates the result analysis, Section "Discussion" gives discussion, and Section "Conclusion" provides the conclusion.

Proposed Method

The network model integrated an application scenario in which multiple mobile devices (MDs) performed N tasks. These include MEC servers that contain required dispensation and storage functionalities are placed at BS or access point (AP). It is assumed that each MD is related to its home BS or AP, the BS or AP with which it has the strongest link. The home BS serves two parts: First, it is responsible for (1) performing tasks if it has enough resources to handle the tasks required second, it will act as middle node passing tasks to other BSs or even servers far away. End users are accessing MEC servers while it accessing other MEC servers in close proximity or other servers at a further distance through back haul means. As the set of MDs, consider $\varphi = \{1, 2, \ldots, Q\}$ and to denote set of tasks by $J = \{1, 2, \ldots, N\}$. Each task is described by three characteristics $J(s_k, c_k, D_k)$, where s_k (in bits) is the input data size and c_k (in MIPS) is the computational intensity or computational workload and D_k (in seconds) is the maximum allowable time delay. These parameters are estimated using task profilers derived from the nature of the application. Further, consider $\Phi = \{1, 2, \ldots, s, \ldots, S\}$ denote group of MEC in a cooperative domain, where total amount of servers is S and each of the server has computation K and the bandwidth capacity B. The amount of the allocated resources within a task k in each MEC server is denoted by $s \in S$. Figure 8.1 depicts the task offloading system model.

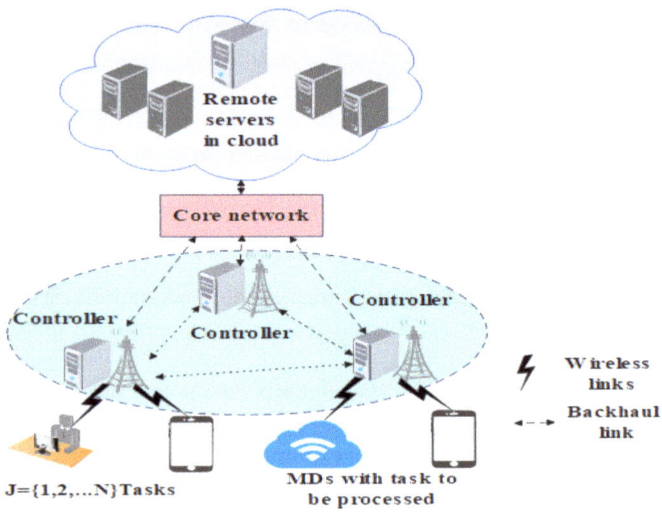

Fig. 8.1 Process of proposed methodology

Communication Model

In this scenario, users transmit service requests to their home BS/AP which is known as home BS. The home BS first asks whether it offers the required service. If the service is not complied with locally, the request is transferred to a neighboring BS or AP to process the request. Let P_η denote the power that is consumed by a user's equipment for the user to communicating with BS and let ψ_η represent channel improvement of user η to the base station. The data rate of user η when sending a request to the BS/AP is determined by Eq. (8.1),

$$R_\eta = B\log_2\left(1 + \frac{P_\eta \psi_\eta}{\sum_{k\in J} x_k P_k \psi_k + \sigma^2}\right) \tag{8.1}$$

Here, the decision variable x_k is binary, which means $x_k \in \{0, 1\}$ when task k is performed in the MD and $x_k = 0$ when it is offload into MEC layer. The symbol $\sum\limits_{k\in J} x_k P_k \psi_k$ describe the interferences that other MD make to the home BS while uploading tasks to MEC servers, while σ^2 is noise power and B is bandwidth. BS is equipped with control server where, different tasks are scheduled and assigned to MEC servers in a way that requires minimum request executive cost.

Service Utility Cost Model

This work limits itself to binary offloading, where an entire application is either solved on the MD or entirely on the MEC servers without partial computation. For local computation, let f_{dev} represent the processing capability of the MD, which is conveyed by the CPU frequency. If no offloading is done, the delay cost is computed as $D_L = \frac{C_k}{f_{\text{dev}}}$ in second; and energy consumption cost as $E_L = \xi d (f_{\text{dev}})^2$, where ξd is a consumed energy. If the required resources for processing a task are available, the home BS processes it task is offloaded to MEC layer. Otherwise, it is delivered to the closest MEC server in cooperative domain to other distant servers if resources permit it. Tasks that are offloaded thus have these two cost implications; communication cost and computation cost. The offloading parameters are determined to $x_k \in \{0, 1\}$, here $x_k = 1$ if k th task is computed locally on the MD and $x_k = 0$ if it is offloaded to the MEC layer. Furthermore, we define $y_k \in \{0, 1\}$, such that $y_k = 0$ means to perform the offloaded task k at the home BS and $y_k = 1$ means the task is to forward the task to one or more nearest MEC servers in cooperative domain or other remote depending on the available resources.

Delay-Cost Model. For instance, where an image taken by an IoT device has to be transferred to capable MEC servers for processing the handoff mechanism is incurred several times. They include uploading, overhead delay and task scheduling, delay for executing those tasks, and the time taken for the server to acknowledge the MD. To simplify the evaluation metrics, the response delay is not considered in this work. We denote the delay that occurred during the transmission of the task from the MD to the MEC server as $D_{d,s}$ and it determined by the formula mentioned as $D_{d,s} = \frac{s_k}{R_\eta}$ where s_k is the size of the task while R_η is the data transmission rate. The overhead delay termed as $D_{qo} = \frac{s_k}{\omega}$ where, ω is the mean rate of subcarrier allocation. Also, $D_{s,j}$ computes the communication delay experienced when one task is transferred from home Bs of s to another adjacent BS j within the cooperative area using X_2 link or to a distant server through a backbone link.

Energy-Cost Model. Consider $F = \{1, 2, \ldots, \omega, \ldots, F\}$ is set available subcarriers each with average B_ω to transfer task k to MEC server. The parameter ω_{ks}^{ω} is the offloading indicator. Here, $\omega_{ks}^{\omega} = 1$ stating that, task k is offloaded to MEC server s with ω subcarrier, else $\omega_{ks}^{\omega} = 0$. The energy consumption is worked out as, $E_{d,s} = P_{d,s} D_{d,s} = P_{d,s} \frac{s_k}{R_\eta}$ where, $P_{d,s}$ is the uplink transmission power, R_η is the rate of uplink data transmission and s_k the size In a similar manner, the energy consumed in communication overheads within the MEC servers is expressed as $E_{s,j} = P_{s,j} \frac{s_k}{R_\eta}$, where, $E_{s,j}$ is a transmission power between the MEC servers Here, $P_{s,j} = 0$, if a task is not forwarded to another server from the present server. In addition, energy cost for tasks offloaded into an approximate specific MEC sever is expressed as E_e which is captured as $E_e = \xi_{\text{ser}} (f_{\text{ser}})^2 c_k$, where ξ_{ser} is a consumed energy of MEC sever.

Task Offloading Strategy

The MOA is a population-based intelligence algorithm (Srinivasan et al. 2021) based on the natural actions of meerkats, members of mongoose family, when they are searching for food. Meerkat clans are usually composed of about twenty members but probably reach fifty or more. These animals hunt for food in groups; one guarding the area against any threats, while the leader is foraging, digging, and hunting. The head of the clan is extremely important because he informs the clan about the finding of food and covers the body with mud to conceal it from the side of the predators. Altogether, when prey is sighted, all members move to a new location as a result of a signal from the chief. They mimic the Meerkats' foraging behavior with swarming and social interactive characteristics. Search leaders adapt their positions according to the interactions which take place to enhance the search. Meerkats born blind are usually fed by their parents, but as they mature, they are able to forage for food on their and create distinguishable large and fast moving groups. Their movement pattern in the early stages is slower and more deliberate than in the later stages and accelerate toward up to 30–48 kmh. This behavior is simulated by the algorithm in the optimization techniques it uses when solving optimization problems. The inertial random displacement of the active sub-population mimics the wide scouting for contingents of solution (e.g., the initial identification of the first power point of the PV panel), while the signals from the leader imitate the narrowing of the range of search to hone the global maxima (e.g., the identification of the MPP of the PV panel). Observations of the social interaction as well as the interactions of the clan members facilitate the process of reaching out toward the best solution. In MOA, the leader position is set as the VMPP, with the best fitness value of PMPP, as a target, while other agents follow, based on social signals, such as barks or whistles. The cooperative movement helps to transfer the entire clan in a systematic and systematic manner to the best solution.

- Step 1: Initialization: Design a random clan of people and set the clan, foraging and care numbers, as well as other characteristics of the environment such as the worst feed and care rates and the rest.
- Step 2: Try to assess the fitness of every person in the community.
- Step 3: In a way, establish which person is the "fittest" for the position and give him the role of the "guard."
- Step 4: Create two distinct groups for the members of the clan; foragers and those who take care of the kids.
- Step 5: Mainly to provide an output consisting of neighboring solutions for the foraging group.
- Step 6: Replace the lowest performers of the foraging group with the highest performers from the care group.
- Step 7: Remove the lowest performing individual in the care group and put in the new randomly generated individual.
- Step 8: If a better individual than the current sentry is found then only change the sentry.

The extended version of the algorithm (Hussein and Yousif 2022) is called the MCA-CC. To the above existing one, this improved version is proposed to enhance searching capability of MCA in searching for good or better solutions. As for the MCA in MCA-CC, the initial solutions are produced by the use of chaotic maps while in the basic MCA, random generation is used. It is implemented to overcome the shortcomings of randomness, which when combined with the determinism part of Chaos increases the exploration aspect of the algorithm greatly. In this paper, the familiar logistic map is employed to interrogate the solutions at the start. The initial solution is generated during the first iteration of the logistic mapping, using a random predetermined value as first iteration of logistic mapping. The logistic map is presented in Eq. (8.2).

$$c(t+1) = \mu c(t)(1 - c(t)) \tag{8.2}$$

where, μ is used to fulfill the chaotic behavior. At each generation of foraging group, using a neighboring search (2-opt) each Meerkat updates position and a portion of the best performing of Meerkats are designated as an elite group. Elite Meerkats work through sentry for this purpose and retrieve new solutions through a unique crossover operation of two points. In this operation, one solution is selected from sentry as solution 1 and one solution from the elite is selected as solution 2. Two distinct break points are picked independently at random, and genes among these two points are copied from first parent; beyond this region the genes are copied from the second parent, thus comprising a child or offspring solution. This process allows the proposed algorithm to use the experience of the clan members to build promising solutions. Further, for the crossover operator the two match a like value from the two parents assuring that the offspring is a combination of the two parents hence improving on the exploration aspect of the algorithm and the ability to always arrive at the global best.

Result Analysis

The MCA-CC performance is simulated in Python with system requirements of windows 10 OS, intel i5 processor and RAM 8 GB. The execution time (s), EC (mJ), and service cost ($) are measured for various no. of tasks. In Table 8.1, the execution time (s) is calculated with various no. of tasks like 20, 40, 60, 80, and 100. The Salp Swarm Algorithm (SSA), Osprey Optimization Algorithm (OOA), and MCA performance are examined and matched with MCA-CC. The MCA-CC obtains 110 s, 250 s, 340 s, 460 s, and 530 s of execution time with various no. of tasks 20–100 correspondingly which is better than SSA, OOA, and MCA.

In Table 8.2, the EC (mJ) is calculated with various no. of tasks like 20, 40, 60, 80, and 100. The SSA, OOA, and MCA performance are examined and matched with MCA-CC. The MCA-CC obtains 10 mJ, 30 mJ, 55 mJ, 70 mJ, and 85 mJ of EC with

Table 8.1 Execution time (s) result

No. of tasks	SSA	OOA	MCA	MCA-CC
20	190	170	150	110
40	310	290	270	250
60	420	380	360	340
80	530	490	470	460
100	600	590	560	530

Table 8.2 Energy consumption (mJ) result

No. of tasks	SSA	OOA	MCA	MCA-CC
20	40	35	15	10
40	70	55	40	30
60	95	80	65	55
80	95	85	80	70
100	120	105	95	85

Table 8.3 Service cost ($) result

No. of tasks	SSA	OOA	MCA	DM-STO
20	35	30	25	20
40	100	95	85	75
60	140	130	120	110
80	180	170	160	150
100	220	210	200	195

various no. of tasks 20–100 correspondingly which is better than SSA, OOA, and MCA.

In Table 8.3, the service cost ($) is calculated with various no. of tasks like 20, 40, 60, 80, and 100. The SSA, OOA, and MCA performance are examined and matched with MCA-CC. The MCA-CC obtains 20$, 75$, 110$, 150$, and 195$ of service cost with various no. of tasks 20–100 correspondingly which is better than SSA, OOA and MCA.

Discussion

The MCA-CC is more effective in noisy or multiple objective environments where the ability to retain the diversity of solutions is of paramount importance. Due to the fact that chaotic maps are adaptive in their nature, the algorithm follows the problem landscape much more flexibly, making it much more robust. Additionally, the crossover strategy improves the convergence rate while at the same time does not affect the quality of solutions obtained and thereby makes MCA-CC is efficient and accurate optimization model. These qualities make MCA-CC a versatile optimization tool for practical problems in MEC systems or when to buy energy to allocate it to resources, if with optimal performance under certain constraints is desirable.

Conclusion

The MCA-CC is proposed in this research for energy-aware computation offloading in edge computing. The MCA-CC enhances the exploration ability of algorithm by including chaotic maps, thus reducing the possibility of early convergence to a local solution. Chaotic maps bring in non-linear randomness into the search space and this enhances the utilization of several solution spaces efficiently. The crossover strategy enhances the exploration abilities and helps to transfer information between candidate solutions that leads to optimize task offloading and resource allocation. Also, crossover is applied with chaotic map making it possible to enhance the level of exploration without compromising on the exploitation and to avoid trapping in exploitation suboptimal areas. The MCA-CC provides better performance in terms of EC, execution time which makes it effective for managing computation offloading in edge computing. The MCA-CC obtains execution time of 530 s, EC of 85 mJ and service cost of 195$ which is better than state-of-art approaches. In future, various no. of tasks are considered to show the effectiveness of proposed approach.

References

Alharbi HA, Aldossary M, Almutairi J, Elgendy IA (2023) Energy-aware and secure task offloading for multi-tier edge-cloud computing systems. Sensors 23(6):3254

Almuseelem W (2023) Energy-efficient and security-aware task offloading for multi-tier edge-cloud computing systems. IEEE Access 11:66428–66439

Baker T, Al Aghbari Z, Khedr AM, Ahmed N, Girija S (2024) EDITORS: energy-aware dynamic task offloading using deep reinforcement transfer learning in SDN-enabled edge nodes. Internet Things 25:101118

He C, Wang R, Tan Z (2020) Energy-aware collaborative computation offloading over mobile edge computation empowered fiber-wireless access networks. IEEE Access 8:24662–24674

Hussein SA, Yousif AY (2022) An improved Meerkat Clan algorithm for solving 0-1 Knapsack problem. Iraqi J Sci 773–784

Li Z, Chang V, Ge J, Pan L, Hu H, Huang B (2021) Energy-aware task offloading with deadline constraint in mobile edge computing. EURASIP J Wirel Commun Netw 2021:56

Mehrabi M, Shen S, Hai Y, Latzko V, Koudouridis GP, Gelabert X, Reisslein M, Fitzek FH (2021) Mobility-and energy-aware cooperative edge offloading for dependent computation tasks. Network 1(2):191–214

Sada AB, Khelloufi A, Naouri A, Ning H, Dhelim S (2024) Energy-aware selective inference task offloading for real-time edge computing applications. IEEE Access 12:72924–72937

Sellami B, Hakiri A, Yahia SB (2022a) Deep reinforcement learning for energy-aware task offloading in join SDN-Blockchain 5G massive IoT edge network. Futur Gener Comput Syst 137:363–379

Sellami B, Hakiri A, Yahia SB, Berthou P (2022b) Energy-aware task scheduling and offloading using deep reinforcement learning in SDN-enabled IoT network. Comput Netw 210:108957

Silva J, Marques ER, Lopes LM, Silva F (2021) Energy-aware adaptive offloading of soft real-time jobs in mobile edge clouds. J Cloud Comput 10(1):38

Srinivasan V, Boopathi CS, Sridhar R (2021) A new Meerkat optimization algorithm based maximum power point tracking for partially shaded photovoltaic system. Ain Shams Eng J 12(4):3791–3802

Suganya B, Gopi R, Kumar AR, Singh G (2024) Dynamic task offloading edge-aware optimization framework for enhanced UAV operations on edge computing platform. Sci Rep 14(1):16383

Zhang H, Liu X, Bian X, Cheng Y, Xiang S (2022) A resource allocation scheme for real-time energy-aware offloading in vehicular networks with MEC. Wirel Commun Mob Comput 2022(1):8138079

Zhao M, Yu JJ, Li WT, Liu D, Yao S, Feng W, She C, Quek TQ (2021) Energy-aware task offloading and resource allocation for time-sensitive services in mobile edge computing systems. IEEE Trans Veh Technol 70(10):10925–10940

Chapter 9
Sine Cosine Learning Factor with Artificial Jellyfish Search Optimization-Based Resource Allocation for Edge Computing in IoT Networks

Sri Shakthi Sarath Chintapalli and Siva Surya Narayana Chintapalli

Abstract The edge computing is a developing architecture that takes storage and computation resources to edge network and generates optimal services and applications near to the end-users. The existing methods are suffered from high energy consumption and delay which leads to reduce the performance. Therefore, this research proposed a Sine Cosine Learning Factor with artificial jellyfish search optimization (SCAJSO)-based resource allocation for edge computing in Internet of Things (IoT). With the proposed SC learning factors, the consideration and learning from both the global optimum agent solutions and the random solutions within the search space offer enhanced learning. This developed learning mechanism increases convergence speed further to facilitate faster resources allocation. SCAJSO also helps to avoid premature convergence and maintain relatively equal contribution of the search space. The SCAJSO achieves throughput of 0.937 Mbps, energy of 42.78 mJ and delay of 5.73 ms for 100 number of tasks which is better than other state-of-art methods.

Keywords Artificial jellyfish search optimization · Edge computing · Internet of Things · Resource allocation · Sine Cosine Learning Factor

S. S. S. Chintapalli (✉)
School of Computing, SRM Institute of Science and Technology, Kattankulathur Campus, Chennai, India
e-mail: shakthisarath20@gmail.com

S. S. N. Chintapalli
Department of Computer Science and Engineering, Birla Institute of Technology, Mesra, Ranchi, India
e-mail: cssuryanarayana@bitmesra.ac.in

G. P. Ramesh et al. (eds.), *Empowering IoT with Distributed Edge Computing*, IoT and Wireless Communication Networks, https://doi.org/10.1007/978-981-96-9455-6_9

Introduction

The emerging 5G and Beyond 5G (B5G) technology's purpose is to improve the Quality of Service (QoS), increase the efficiency and increase the Internet of Things (IoT) network capacity (Seid et al. 2021). These advances are anticipated to reduce computation costs by utilizing state-of-art network services and learning algorithms. Using IoT with machine learning gives a revolutionary shift that increases future network technology and optimizes the IoT across multiple industries (Dong et al. 2021). Modern smart applications are based on deep neural network (DNN) architectures consisting of minimal processing elements (PEs) and matrixes in series and in cascade. These computations are enabled into IoT device control by the light-weight DNN models (Gong et al. 2022). But the large number of Industrial IoT devices and computing requirements of complex applications like augmented reality, real-time online gaming and ultra-high-definition streaming brings large data processing, architectural complexity, and resource management problems (Aghapour et al. 2023). Exacerbating these issues are IoT devices operating in far or difficult to reach sites for example, underwater, forests, deserts, or highways (Zang and Wang 2023). In such regions, terrestrial base stations (BSs) are scarce and installation of more BSs is unprofitable. In order to cope with this demand, unmanned aerial vehicles (UAV) systems with edge servers have appeared as perspective solutions (Liu et al. 2024). Self-flying computers also referred to as UAVs are mobile and versatile, they easily leveraged to perform computation services where they needed at any given time (Tianqing et al. 2021). Mobile edge computing simplifies these efforts additionally by bringing cloud services closer to the user while also providing efficient service provision (Sharif et al. 2021). The more complex solution of decentralization of decision-making means deploying it onto the edge host or end device that, in turn, controls the distribution of its resources according to its needs. This approach speeds up adaptation and optimizes usage of the limited resources, laying foundation for reliable, large-scale IoT networks in various and difficult conditions (Mahmood et al. 2022).

Wang and Wang (2021) suggested a resource allocation optimization strategy in cloud-edge computing. Initially, heterogeneous 5G cloud-edge computing was established. Furthermore, according to the status of the network, the computing intensities of devices are divided to local processing and edge computing. The delay in computing, the communication and the provision of computing resources of edge servers were modeled. Then, using delay and energy consumption of computing resource allocation as the optimization target, significance was assigned for the subtasks to provide the best allocation of the computing resources.

Peng et al. (2022) developed a computation offloading and resource allocation by edge-cloud computing. The computation offloading and resource allocation were modeled as multi-objective problem and end-edge-cloud cooperative optimization was developed. To assess the viability and performance benefits of developed model in energy and time cost of IIoT devices and resource utilization and load distribution of edge servers, extensive experiments and performance analyses were also performed.

Tan et al. (2021) introduced an energy-efficient task offloading and resource allocation for edge computing. The delay-sensitive tasks of users were determined to be computed either locally, on collaborating devices or MEC servers. The overall goal was confined to reducing the total energy consumed by all mobile users and the constraint was on delay. This was modeled as mixed-integer nonlinear programming (MINLP) model that considers decisions on task offloading and collaboration, subcarrier and power allocation and computing resource provisioning. As a result of formulating the problem as an MINLP, an alternation method framework at two levels was suggested.

Ke et al. (2021) presented a computation offloading and resource allocation by deep reinforcement learning in mobile edge computing. A decentralized optimization system for partial computation offloading and resource allocation using deep reinforcement learning (DOCRRL) was developed for edge computing. The channels are time-varying, computation workloads are arriving over time in random manner and SINR, optimum policy for decision-making has been learned by the DOCRRL algorithm under latency and risk constraints. This approach addressed efficiency of dimensionality which is associated with the action and state spaces.

Huang et al. (2021) presented a dynamic resource allocation in mobile edge computing with multi-user and multi-server. The mobile user behavior was modeled in terms of an evolutionary game and deterministic and stochastic models where built to describe the evolution of the mobile users and the evolutionary equilibrium is the solution. To get the evolutionary equilibrium, an evolutionary algorithm was developed. In addition, based on the noncooperative game, competition between edge cloud servers (ECSs) was investigated and the iterative approach was provided to compute for Nash equilibrium. This enabled ECSs to vary the quantity of resources that are employed and the price that is charged for mobile user services so as to elicit more users.

The contributions are given as follows:

- The utilization of SC learning factor in the AJSO takes numerous benefits such as adaptability, resource utilization, and reduced delay for optimized resource management of edge computing in IoT networks.
- These learning factors improve the performance of the balancing of exploration and exploitation in the algorithm since jellyfish learning is based on random solutions and the best candidate solutions in the search space.
- This makes the dual-learning mechanism an effective process for accelerating the system to converge to near-optimal solutions at different resource allocation situations.

The research organization is arranged as: Section "Proposed Method" explains about the proposed method, Section "Result Analysis" discussed about the result analysis, Section "Discussion" details the discussion about proposed method, and Section "Conclusion" explains the conclusion.

Proposed Method

IoT network topology consisting of multiple end devices (IoT devices) and one gateway (edge device). The gateway extracts information from end devices in its range and then applies analysis on it with its built-in edge server. Every end device is constantly computing numerous computation tasks, but it has a very limited computational capability and power. Such tasks are delegated to the gateway to improve performance through consumptions of less power and also shorter times to complete tasks. The time line is discretized into epochs each spanning a timescale of η, with epoch k labeled as $0 < k \leq K$ where K is a total no. of epochs. The set of end devices in the network is represented by $U = \{u_1, \ldots, u_U\}$. The channel gain state transition probabilities are described using a finite state discrete time Markov chain. Many individual computational instructions are performed at each end device and each of a varying size in terms of CPU cycle requirements. End-device task queue is depicted as $T = \{T_1, \ldots, T_{\max}\}$, where $T_{\max}$ is the maximum number of tasks in the queue. The inter-arrival times are denoted by $T = \{0, 1\}$ imply that it has a task at that epoch and this is accompanied by a randomly chosen $M = \{m_1, \ldots, m_M\}$ and $T = 1$ imply that no arrival has occurred in the epoch. As depicted in Fig. 9.1, computation tasks performed at end-device or they offloaded to gateway performed at edge server.

In every IoT, end devices perform part of computation within the epoch of time, while others move the computation to the gateway. To achieve this, at start of every

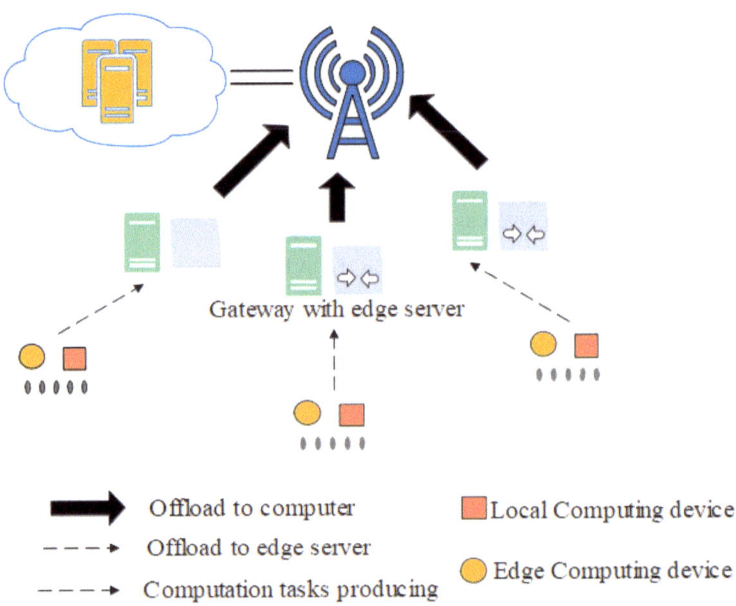

Fig. 9.1 Computation task offloading in IoT network

epoch k, every end device makes an independent decision of offloading, $O = \{1\} \cup \{0\} \cup \{-1\}$ and chooses the amount, $P_t = \{P_1^k, \ldots, P_{\max}^k\}$ if it chooses to offload its task to the edge device. If end device not select to task offloading, cost consists of local computation power consumption and completion time, while the transmit power is set at $P_t^k = 0$. When $O^k = 1$, this means that for that particular task the offload has been chosen by the end device and the transmit power is selected from the power set $P_t^k \in P_t$. In both cases, the task is said to have been completed with success, however, if there is a disruption to the transmission between end device and gateway, task failed and $O^k = \{-1\}$.

Local Computing Mode

In local computing mode, all computations required in the completion of the task are carried out on the local machine and is represented by $O^k = 0$. For the purpose of the model, we presume that the edge server grants each of the end devices a fixed and equal amount of CPU resources that accomplish the computation tasks on time.

Let f_d is the fixed value of the CPU of an end device defined in cycles per second. This is parameter which represents no. of CPU cycles required to develop one bit of data and P_d is a per CPU cycle power consumption. The energy profile of local device is thus the energy per bit of data transmitted or $f_d P_d$. For a given time, epoch k, power consumption for one computational task is stated as Eq. (9.1),

$$P_{cd}^k = f_d P_d m^k \tag{9.1}$$

where, D_d is the computational capability or computational ability of end device, that is a number of CPUs end device performs per second of time. The CPU resources at end device during epoch k are given as a set of percentages. The combinatorial set of resultant resistances $R^k = \{r_{d1}, r_{d2}, \ldots, 1\}$. The local computing latency L_d^k is calculated as Eq. (9.2),

$$L_d^k = \left(f_d m^k\right)/D_d \tag{9.2}$$

However, a major research issue in edge computing networks is the energy efficiency versus the execution delay of tasks. It is impossible to decrease both of them at the same time, therefore the goal is not to avoid increasing one of them, that would lead to increasing the other one. To achieve this, cost function for local computing mode is defined as Eq. (9.3),

$$C_{\text{loc}}^k = P_{cd}^k + \beta L_d^k \tag{9.3}$$

where, β is the weight factor of the game which balances consumption power of the device with task delay.

Offloading Computing Mode

In this mode, end devices use a protocol known as the time division multiple access (TDMA) in order to transmit data to gateway. This makes interference between devices small during their time slot in epoch k within which it transmits. We use g^k to signify channel gain from an end device to gateway and use a subscript f to denote fact that this gain is fixed during the period of offloading. The transmit power of the device is denoted as P_t^k. Consequently, the achievable data transmission rate (in bits per second) is expressed as Eq. (9.4),

$$R_k = B_\omega log_2\left(1 + \frac{P_t^k g^k}{\sigma^2}\right) \tag{9.4}$$

where, B_ω is the bandwidth and σ^2 is the variability associated with Additive White Gaussian Noise (AWGN). The edge server's computational capability received by each device is denoted as D_s. The computational power of edge server for processing offloaded data is computed as Eq. (9.5),

$$P_{cs}^k = f_s P_s m^k \tag{9.5}$$

where, m^k represents the no. of CPU cycles which was required. The latency for edge computation as Eq. (9.6),

$$L_s^k = \left(f_s M^k\right)/D_s \tag{9.6}$$

By combining the transmission and computation costs, the total cost function for offloading computing mode is defined as Eq. (9.7),

$$C_{\text{off}}^k = P_{cd}^k + P_t^k + \beta\left(L_s^k + L_t^k\right) \tag{9.7}$$

where β is the weighting of electrical power usage and delay.

Resource Allocation

The AJSO (Siddiqui et al. 2021) is a novel metaheuristic algorithm which is based on the movement of jellyfish in the ocean. In resource allocation, the AJSO works through simulating two behaviors active and passive movements which begins with initial population of solution which provides better allocations. The passive movements explore global search space through distributing solutions which ensures diversity and reduce the premature convergence. The balance between these movements are maintained during the search process. This adaptability enables AJSO to effectively allocate resources in complex and dynamic systems which achieves optimal or near-optimal solutions while reducing energy consumption. The resource allocation is made depending on the type of movement such as passive like in the event influenced by ocean currents or swarming; active like the movements of the specific jellyfish. There is a special time control that determines when one movement mode changes to the other. First, it searches for food swims in the sea and it is fond of areas with high concentration of food. Within each iteration, the algorithm analyses and ranks all the food quantities and finally set the best global position α^* for the highest concentration of food. The first and the final position of jellyfish is given by the value $f(\alpha)$, while the position where food is most plentiful is represented by the current best as α. The jellyfish exhibit two types of motion: active and passive. The active motion is distinguished where the swarm has only just begun to emerge, where nodes are still connecting to one another. In this phase, all jellyfish change their positions with respect to their current positions and their new positions are calculated by Eq. (9.8),

$$\alpha_L^{t+1} = \alpha_L^t + \text{rand}(0, 1) \times \text{rand}(0, 1) \times \mu \tag{9.8}$$

where mean of all positions of jellyfish is denoted by μ. The passive motion is assigned arbitrarily when a position and velocity vector of a jellyfish L is correctly chosen to control another jellyfish by M for its selection of new position. If the quantity of food at M location is larger than that at location L shifts its position toward M. Therefore, if L was to carrying more food than it chases off from this position in order to search for other possible positions. This behavior enables jellyfish to move searching for the best vantage point and thus to get the best out of what is available to it. From these directional movements, the global best position is fine-tuned until its algorithm locks into the best possible solution as Eq. (9.9). The new position of jellyfish is given in Eq. (9.10),

$$\text{Direction} = \begin{cases} \alpha_M^t(t) - \alpha_L^M(t) \text{ if } f(\alpha_L) \geq f(\alpha_M) \\ \alpha_L(t) - \alpha_M(t) \ \text{ if } f(\alpha_L) < f(\alpha_M) \end{cases} \tag{9.9}$$

$$\alpha_L^{t+1} = \alpha_L^t + \text{rand}(0, 1) \times \text{Direction} \tag{9.10}$$

where, f is a harmonic reduction index at α location. The time control technique is used to control as well as quantify the type of jellyfish movement over time. It decides whether jellyfish are affected by the ocean current or else swims in swarm. This kind of technique utilizes a constant C_0 as well as a time control function t_c which will random value that float between 0 and 1. The C_0 as fixed to 0. 5 is a midway point between 0 and 1 which represents the mean of all the values. The time control function is expressed mathematically by Eq. (9.11) which include the current iteration, t_0 and the maximum number of iterations, Max_{iter}.

$$t_c = \left| \left(1 - \frac{t}{\text{Max}_{\text{iter}}} \right) \times (2 \times \text{rand}(0, 1) - 1) \right| \tag{9.11}$$

The jellyfish are often found to swim in a pattern following the availability of food so as to from a swarm. Jellyfish may move within the swarm to another ocean current; new swarms come from the variation in factors such as temperature or wind direction. If $t_c \geq 0.5$, the movements of jellyfish are directed to the ocean currant shown by Eq. (9.12),

$$\text{current} = \alpha^* - \beta \times \text{rand}(0, 1) \times \mu \tag{9.12}$$

where β is distribution coefficient and μ is the mean position of all identified jellyfish. If $\text{rand}(0, 1) < (1 - t_c)$, jellyfish move in swarm based and in either active or passive manner. Namely in particular, when $\text{rand}(0, 1)(1 - t_c)$ passive movement occurs and otherwise active movement is executed. Based on this, the proposed dual movement strategy helps to increase the convergence speed and at the same time avoids the algorithm being stuck at local optima. Thus, what is initiated as diverse population is improved by the means of logistic map, which makes it difficult for the algorithm to converge too soon and which also introduces substantial chaos to the population. The expression of the logistic map is as follows, the dynamic described using the following Eq. (9.13),

$$\alpha_{L+1} = 4\alpha_L(1 - \alpha_L), \quad 0 \leq \alpha_0 \leq 1 \tag{9.13}$$

In each iteration, the feasibility of the boundary conditions is determined and the food quantity at new locations is assessed. The positions of the jellyfish update in sequence until the termination condition, maximum iterations counter Max_{iter} is fulfilled. It continues until $t > \text{Max}_{\text{iter}}$ and the best position in the global, where high amount of food (α^*) is located.

Sine Cosine Learning Factors. The specific motion of jellyfish in exploration phase is Type B in the swarm. Here, replacing the old position of a jellyfish is done based on a certain learning from another jellyfish randomly chosen. However, this random learning process brings several relative restrictions which are as follows: (i) information exchange is ineffective in the population, (ii) disturbing the optimal candidate solution. Moreover, it cause a random change which will slow down the convergence speed. To overcome the above limitations, sine and cosine learning factors ω_1 and ω_2 are incorporated. These factors allow jellyfish to not only learn from any random candidate in the population but to learn from the best candidate within the search range. This dual-learning strategy reduces uncertainty in the exploration phase by learning from the feedback signal and improving the solution space to converge faster and better-quality candidates as Eqs. (9.14) and (9.15),

$$\omega_1 = 2 \cdot \sin\left[\left(1 - \frac{t}{T}\right) \cdot \frac{\pi}{2}\right] \tag{9.14}$$

$$\omega_2 = 2 \cdot \cos\left[\left(1 - \frac{t}{T}\right) \cdot \frac{\pi}{2}\right] \tag{9.15}$$

In Type B movement, when the location update is carried out, the mobility of the jellyfish described by the following Eq. (9.16),

$$P_i(t+1) = \omega_1 \cdot \left(P_i(t) + \overrightarrow{\text{step}}\right) + \omega_2 \cdot \left(P^* + P_i(t)\right) \tag{9.16}$$

where $\overrightarrow{\text{step}}$ is a step vector. The traditional algorithm was based on the random learning only, meaning that when jellyfish learned from some people with certain low fitness values, the convergence was restrained. With the help of sine and cosine learning factor involved, the jellyfish provide a dynamic regulation of exploration–exploitation. This approach ensures they get good solutions from the random search and also get an optimal solution within the search range hence improving the solution quality and convergence rate.

Result Analysis

The SCAJSO performance is simulated in Python with system configuration of windows 10 OS, RAM 8 GB, and intel i5 processor. The throughput, energy, task scheduling rate, and delay are considered for calculating the SCAJSO performance. Table 9.1 describes the throughput performance for SCAJSO performance with different no. of tasks. The Cat Swarm Optimization (CSO), Rat Swarm Algorithm (RSA), and AHSO performance are examined and compared with SCAJSO. The SCAJSO achieved better throughput of 0.984, 0.973, 0.962, 0.945, and 0.937 Mbps which is better than CSO, RSA, and AJSO.

Table 9.2 describes the energy consumption performance for SCAJSO performance with different no. of tasks. The CSO, RSA, and AHSO performance are examined and compared with SCAJSO. The SCAJSO achieved better energy consumption of 29.23, 32.16, 35.62, 38.51, and 42.78 mJ which is better than CSO, RSA, and AJSO.

Table 9.3 describes the task scheduling rate performance for SCAJSO performance with different no. of tasks. The CSO, RSA, and AHSO performance are examined and compared with SCAJSO. The SCAJSO achieved better task scheduling rate of 0.97, 0.95, 0.93, 0.92, and 0.89 which is better than CSO, RSA, and AJSO.

Table 9.1 Throughput (Mbps) for SCAJSO performance

No. of tasks	CSO	RSA	AJSO	SCAJSO
20	0.930	0.950	0.974	0.984
40	0.922	0.938	0.951	0.973
60	0.915	0.930	0.947	0.962
80	0.906	0.925	0.935	0.945
100	0.891	0.904	0.923	0.937

Table 9.2 Energy consumption (mJ) for SCAJSO performance

No. of tasks	CSO	RSA	AJSO	SCAJSO
20	40.12	36.14	32.18	29.23
40	43.39	37.85	35.27	32.16
60	46.25	42.37	38.44	35.62
80	48.63	46.18	41.72	38.51
100	52.47	49.32	45.83	42.78

Table 9.3 Task scheduling rate for SCAJSO performance

No. of tasks	CSO	RSA	AJSO	SCAJSO
20	0.88	0.92	0.94	0.97
40	0.84	0.90	0.89	0.95
60	0.78	0.89	0.88	0.93
80	0.76	0.87	0.87	0.92
100	0.73	0.85	0.83	0.89

Table 9.4 Delay (ms) for SCAJSO performance

No. of tasks	CSO	RSA	AJSO	SCAJSO
20	5.92	4.83	3.58	2.67
40	6.59	5.49	4.73	3.24
60	7.72	6.78	5.21	3.76
80	8.53	7.69	6.51	4.43
100	9.86	8.26	7.68	5.73

Table 9.4 describes the delay performance for SCAJSO performance with different no. of tasks. The CSO, RSA, and AHSO performance are examined and compared with SCAJSO. The SCAJSO achieved better delay of 2.67, 3.24, 3.76, 4.43, and 5.73 ms which is better than CSO, RSA, and AJSO.

Discussion

In edge computing situations when the demand and consumption of resources are not constant and computational requirements are high, the SC learning factor allows to do the same with better precision. By considering these factors, the algorithm optimizes the use of the edge servers, reduces computational delay and power-consuming. Also, the method includes diversity to the search space which makes the approach highly efficient in dealing with diverse IoT. Firstly, the presented design of the AJSO enables a scalable and efficient solution to edge computing in IoT networks. In edge computing scenarios where resource availability is limited by both deployments, SCAJSO flexibility in resource assignment ensures that resources are optimally allocated to reduce delay and enhance quality of the delivered service. This developed learning mechanism increases convergence speed further to facilitate faster resources allocation. SCAJSO also helps to avoid premature convergence and maintain relatively equal contribution of the search space.

Conclusion

The SCAJSO algorithm is proposed in this research for resource allocation in edge-cloud environment. The utilization of SC learning factor in the AJSO considers numerous benefits for optimized resource management in edge computing in IoT networks. These learning factors enhance the performance of balancing exploration and exploitation in the algorithm since jellyfish learning is based on random solutions and the best candidate solutions in the search space. This makes the dual-learning mechanism is an effective process for quickening the system to converge to near-optimal solutions at various resource allocation situations. By considering different computing modes, the algorithm optimizes the use of the edge servers, reduces computational delay, and power-consuming. The SCAJSO achieves throughput of 0.937 Mbps, energy of 42.78 mJ, and delay of 5.73 ms for 100 number of tasks which is better than other state-of-art methods. In future, reinforcement learning will applied to further enhance the resource allocation performance.

References

Aghapour Z, Sharifian S, Taheri H (2023) Task offloading and resource allocation algorithm based on deep reinforcement learning for distributed AI execution tasks in IoT edge computing environments. Comput Netw 223:109577

Dong C, Hu S, Chen X, Wen W (2021) Joint optimization with DNN partitioning and resource allocation in mobile edge computing. IEEE Trans Netw Serv Manage 18(4):3973–3986

Gong Y, Yao H, Wang J, Li M, Guo S (2022) Edge intelligence-driven joint offloading and resource allocation for future 6G industrial Internet of Things. IEEE Trans Netw Sci Eng 11(6):5644–5655

Huang X, Zhang W, Yang J, Yang L, Yeo CK (2021) Market-based dynamic resource allocation in mobile edge computing systems with multi-server and multi-user. Comput Commun 165:43–52

Ke HC, Wang H, Zhao HW, Sun WJ (2021) Deep reinforcement learning-based computation offloading and resource allocation in security-aware mobile edge computing. Wireless Netw 27(5):3357–3373

Liu Q, Mo R, Xu X, Ma X (2024) Multi-objective resource allocation in mobile edge computing using PAES for Internet of Things. Wireless Netw 30(5):3533–3545

Mahmood OA, Abdellah AR, Muthanna A, Koucheryavy A (2022) Distributed edge computing for resource allocation in smart cities based on the IoT. Information 13(7):328

Peng K, Huang H, Zhao B, Jolfaei A, Xu X, Bilal M (2022) Intelligent computation offloading and resource allocation in IIoT with end-edge-cloud computing using NSGA-III. IEEE Trans Netw Sci Eng 10(5):3032–3046

Seid AM, Boateng GO, Mareri B, Sun G, Jiang W (2021) Multi-agent DRL for task offloading and resource allocation in multi-UAV enabled IoT edge network. IEEE Trans Netw Serv Manage 18(4):4531–4547

Sharif Z, Jung LT, Razzak I, Alazab M (2021) Adaptive and priority-based resource allocation for efficient resources utilization in mobile-edge computing. IEEE Internet Things J 10(4):3079–3093

Siddiqui NI, Alam A, Quayyoom L, Sarwar A, Tariq M, Vahedi H, Ahmad S, Mohamed ASN (2021) Artificial jellyfish search algorithm-based selective harmonic elimination in a cascaded h-bridge multilevel inverter. Electronics 10(19):2402

Tan L, Kuang Z, Zhao L, Liu A (2021) Energy-efficient joint task offloading and resource allocation in OFDMA-based collaborative edge computing. IEEE Trans Wireless Commun 21(3):1960–1972

Tianqing Z, Zhou W, Ye D, Cheng Z, Li J (2021) Resource allocation in IoT edge computing via concurrent federated reinforcement learning. IEEE Internet Things J 9(2):1414–1426

Wang J, Wang L (2021) A computing resource allocation optimization strategy for massive internet of health things devices considering privacy protection in cloud edge computing environment. J Grid Comput 19(2):17

Zhang X, Wang Y (2023) DeepMECagent: multi-agent computing resource allocation for UAV-assisted mobile edge computing in distributed IoT system. Appl Intell 53(1):1180–1191

Chapter 10
Active Fitness-Based Al-Biruni Earth Radius Optimization Algorithm to Secure Edge Computing for Internet of Things-Based Smart Cities

Siva Surya Narayana Chintapalli, Satya Prakash Singh, and Vijaya Lakshmi Sarraju

Abstract The technology advancement has contributing huge resources for the development of smart cities. The Internet of Things (IoT)-based appliances are rising rapidly for its vast benefits. The edge computing stores the data transmitted by the IoT devices and also provides the security to the stored data. The IoT data is transmitted to the edged computing for the data storage and security that maintains the privacy of the users. The existing methods tried to improve the data transmission security but did not reach the expected outcome because they fail to resist for multiple vulnerable attacks in the network. To solve the existing problem the Active Fitness-based Al-Biruni Earth Radius (AFBER) optimization model is proposed to improve the security by prioritizing the edge computing tasks. The performance of the developed AFBER model is evaluated in terms of throughput and energy consumption. The incorporation of active fitness in the Al-Biruni Earth Radius (BER) method has advanced searching ability of the model for identifying the solution for data transmission. The AFBER methods has gained higher throughput of 46.52 Kbps and minimal energy consumption of 0.09 J for 25 number of nodes compared to the existing Intelligent Buffalo-based Secure Edge-enabled Computing (IB-SEC) algorithm.

Keywords Active fitness-based Al-Biruni earth radius optimization algorithm · Data transmission · Edge computing · Internet of Things · Security · Smart cities · Trust

---

S. S. N. Chintapalli · S. P. Singh · V. L. Sarraju (✉)
Department of Computer Science and Engineering, Birla Institute of Technology, Mesra, Ranchi, India
e-mail: vijayalakshmi@bitmesra.ac.in

S. S. N. Chintapalli
e-mail: cssuryanarayana@bitmesra.ac.in

S. P. Singh
e-mail: sp.singh@bitmesra.ac.in

G. P. Ramesh et al. (eds.), *Empowering IoT with Distributed Edge Computing*, IoT and Wireless Communication Networks, https://doi.org/10.1007/978-981-96-9455-6_10

Introduction

The scattering of the computing models which transfers the data storage and processing nearer to the devices that produce the data and use it is known as edge computing (Shu et al. 2024). The computing resources are brought together to the edges where the data get processed (Ahmed et al. 2022a). The physical substances which get embedded with the software, sensor, and other required materials are generally known as Internet of Things (IoT) (Ahmed et al. 2022b). The rise in the usage of IoT devices is rising rapidly due to increase in the population. The IoT has occupied our life, without these gadgets we are not able to lead the flexible life (Javeed et al. 2024). The security of the data transmitted by the IoT devices is affected due to multiple vulnerable activity that destroyed the privacy of the stored data (Huang 2024). The edge computing is interconnected with the IoT in order to manage the huge amount of data efficiently with security (Khan et al. 2024). As the IoT devices are transmitting the data continuously, the edge computing helps to filter the large amount of data that minimizes the bandwidth required for the data transmission (Zhou et al. 2022). The delicate data information is frequently managed by the IoT secure the privacy of the data during the process of transmission due to improved cyber security (Liu et al. 2022). To stretch the functionalities of the industries, the IoT is deployed for the adjustment and adaptability of variations in the operational region (Qayyum et al. 2022). As the innovative materials are rising in the current generation, the security of the data obtained from the IoT devices is increasing for the smart city management purpose (Saba et al. 2023). The unauthorized access in the network delayed the user verification that affected the security of IoT data. The contribution of the proposed AFBER method for secure edge computing for IoT-based smart city is given as follows:

- The transmission of data from IoT-based smart city devices to edge computing is optimized to secure the privacy of the data against vulnerable activities to maintain the user trust by prioritizing the tasks.
- The Al-Biruni Earth Radius (BER) optimization model is used to find the optimal solution for balancing the data transmission in the edge computing that increases the security of the data.
- The active fitness is incorporated in the BER forming Active Fitness Al-Biruni Earth Radius (AFBER) algorithm that improved the space search ability of the that advanced the data transmission security to gain the trust of users.

The rest of the section present in this research paper is formatted as given below: Section "Literature Survey" indicates the literature survey. Section "Proposed Methodology" provides the proposed AFBER methodology. Section "Experimental Results" exhibits the experimental results. Section "Discussion" comprises the discussion and Section "Conclusion" gives the conclusion.

Literature Survey

The current section briefs the details of the existing techniques used for secure edge computing for IoT-based smart cities with their benefits and drawbacks. Ali et al. (2022) developed a trust framework for Multiclass Edge Computing (MEC) using prediction model. The Zero Trust Security (ZTS) was incorporated with the MEC that increased the verification process of edge computing users by reducing the energy consumption. The model predicted the vulnerabilities in the network and protected the data against it that increased the security level of the edge computing. However, the model showed poor performance due to lack of optimal solution for the authentication process that decreased the model performance. Latif et al. (2022) presented a trust-based management system for edge computing for IoT devices. The multi-criteria decision analysis (MCDA) was used to select the optimal solution for the to increase the user trust. A lightweighted trust management mechanism was incorporated to maintain the service trust in the edge computing that advanced the security performance of the model in the edge computing. However, the model failed to maintain the stability of the model that affected the security quality of the edge computing. Ali and Khan (2023) implemented a sustainable smart city using IoT with trust-based security. The artificial neural network (ANN) was used with the support vector machine (SVM) to predict the traffic in the network and managed secured the data transmission. This hybrid model advanced the performance of securing the data with minimum energy consumption that improved the model stability. However, the model did not handle the multiple tasks in the network that minimized the security of the data transmission.

Ajao and Apeh (2023) developed a secured edge computing for smart city sustainability based on reinforcement learning with Petri Net model. The genetic algorithm was used for the optimizing the performance of cloud security with secure trust-aware philosopher privacy and authentication (STAPPA) that advanced the security of IoT data transmission. However, the unauthorized access in the network led to latency in the user request that suppresses the stability of the securing the edge computing data for multiple attacks. Irshad et al. (2023) presented an Intelligent Buffalo-based Secure Edge-enabled Computing (IB-SEC) model for the heterogeneous IoT network in the smart cities. The African Buffalo Optimization (ABO) model was used for optimizing the security functioning of the IoT that improved the performance of model to get appropriate and secure data transmission in the edge computing that advanced the reliability of the model. However, the security of the cloud data was affected due to delay in the data verification process that affected the performance of the model. The existing models struggled with many drawbacks during the process of optimizing the data transmission in the IoT and edge computing. The traditional model failed to handle the multiple access in the network that suppressed the performance of data transmission in the edge computing. The vulnerable activities during the process of transmitting the data cause interruption where some of the data packets get lost that suppress the security of the IoT-based smart cities' data. Maintenance of the user trust toward the edge computing is affected by the latency in the data transmission.

Proposed Methodology

The AFBER algorithm is proposed for improving data transmission trust and secure edge computing for IoT smart device. The source, IoT-based smart cities, data transmission are the components used here by the proposed AFBER model for evaluating the security and reliability of the data transmission using MAC layer. The AFBER algorithm optimizes the data packets during the process of transmission so that ensure there is no loss of packets while transmitting the data from the IoT-based smart city devices to the edge computing. Figure 10.1 exhibits the block representation of the proposed AFBER optimization algorithm.

The overview of the proposed AFBER model flow is described here as follows:

- Source: This provides the relationship between transmitter that forward data to various nodes in the infrastructure od smart city.
- IoT-based smart cities: This indicates the different organizations, constructions, sectors that relay of the wireless network where the urban atmosphere get connected with each other with the transmission of data.
- Data transmission: This is responsible for the management of the data flow in smart city in-between the destination and sources.

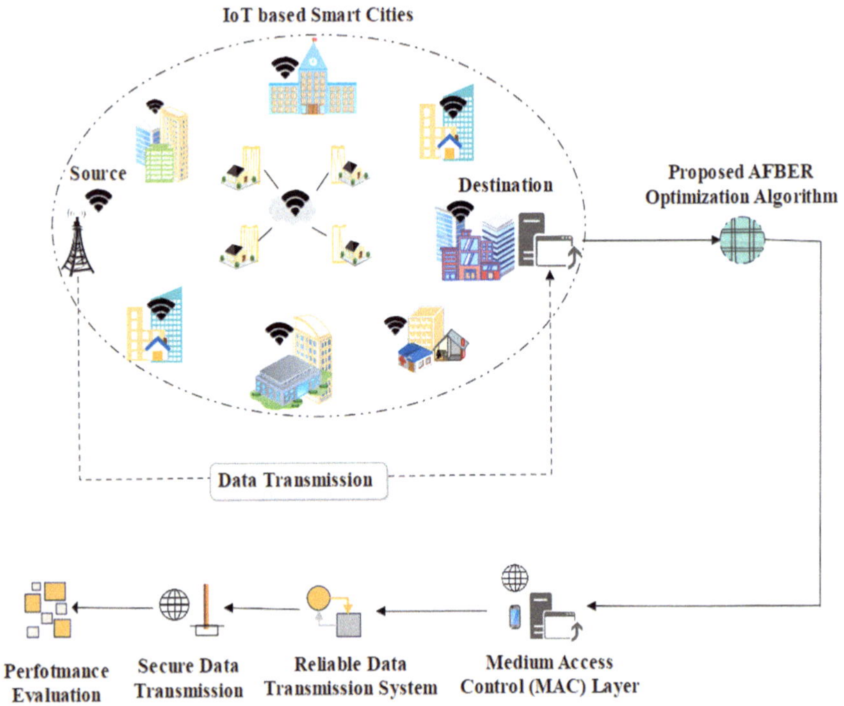

Fig. 10.1 Block representation of the proposed AFBER optimization algorithm

- Proposed AFBER optimization algorithm: The IoT devices data of smart cities is transmitted to the edge computing without any interruption. The model increases the security of the data transmission to gain the trust of the users.
- Medium access control (MAC): This provides the group of protocol that ensures the transmission of data packets reliability and security. The performance of the data transmission is evaluated to observe the functioning of the secured IoT-based smart cities data transmission.

Proposed AFBER Optimization Algorithm

The BER is the optimization algorithm used for the advancement of secured data transmission. The model works on the principle of simulating swarm where the population are gathered for the process of exploration and exploitation. The agents fitness get explore and exploit that improves the agents number in the groups. The BER population is indicated as $S = S_1, S_2, \ldots, S_d \in R$ where d represents the feature. The F exhibits the objective function to evaluate the functioning of agent. Further the agents get optimized in order to identify the finest agent. The fitness function is utilized by the BER model to optimize the performance of agent and determines the size of population. The introverted explorer in the cluster employ tactic to detect the possible new provinces to explore the area where agents are positioned in order to go closer to the finest optimal value. Further, to complete this objective, the abundant possibilities are offered in the neighboring area and take the substitute that is greater to the others with respect to the outcome. The mathematical representation of the BER is given below in Eq. (10.1),

$$S(t+1) = S(t) + D(2r_2 - 1) \tag{10.1}$$

Here $S(t)$ exhibits the solution of t iteration, D is the diameter of the search space. The set answerable in order to discover path to provide the explanations that are previously available in area that are more active. The BER is regulated at the conclusion of every turn where the individual agent have attained the uppermost levels of fitness. This model finishes its task of exploitation on the basis of two discrete tactics. The optimal value gets closer using Eq. (10.2),

$$S(t+1) = r^2(S(t) + D) = r_3(L(t) - S(t)) \tag{10.2}$$

Here $= h\frac{\cos(x)}{1-\cos(x)}$, r_3 provides the random vector, $L(t)$ indicates the best solution. Exploratory the area exposes that the area adjacent the best solution that provides most stimulating potential explanation. Therefore, numerous agents hunt for ways to better conditions by anticipating substitute choices which are moderately similar with the best option. The mathematical representation of the exploratory stage is given in Eq. (10.3),

$$S' = (t+1) = r\big(S^*(t) + k\big), \text{ where } k = 1 + \frac{2 \times t^2}{\text{Max}_{\text{iter}}^2} \tag{10.3}$$

Here $S^*(t)$ gives the best solution, The level of the optimal fitness is obtained based on the previous two iteration value and the outcome bet modified using Eq. (10.4),

$$S(t+1) = k * z^2 - h\frac{\cos(x)}{1 - \cos(x)} \tag{10.4}$$

Here z provides the random value with range [0, 1]. To deliver the uppermost probable level of quality, the BER model finds a substitute solution for the succeeding cycle. The usefulness of superiority accelerates the process of multimodal procedure meets one another. Based on the utilization of mutational strategy and thoroughly evaluating the agent population of the exploration group, this model exhibits the great proficiencies for mineral exploration. The BER technique iteratively explorations for the finest reply and robotically adapts the size of each agent cluster. The responses of the BER get rearranged that ensures agent varied and maintain adequate depth. In the succeeding cycle, a solution that was established by the exploration cluster passed on to the exploitation cluster. The leader in the BER cannot be replaced due to its restricted optimizing ability. To solve this problem, a fitness is developed in the BER model with upgrades the position of the agent based on fitness value. The BER model fitness is estimated using Eqs. (10.5)–(10.7),

$$F_{L_1} = \frac{F_{L_1}}{F_{L_1} + F_{L_2} + F_{L_3}} \tag{10.5}$$

$$F_{L_2} = \frac{F_{L_2}}{F_{L_1} + F_{L_2} + F_{L_3}} \tag{10.6}$$

$$F_{L_3} = \frac{F_{L_3}}{F_{L_1} + F_{L_2} + F_{L_3}} \tag{10.7}$$

Here, $F_{L_1}, F_{L_2}, F_{L_3}$ provide the fitness value. The BER values are updated in the developed AFBER model on the basis of fitness function which is shown in Eq. (10.8),

$$S(t+1) = r^2(S(t) + D) \tag{10.8}$$

The distance vector gets altered based on the calculated values of $F_{L_1}, F_{L_2}, F_{L_3}$ function using Eq. (10.9),

$$D = r_3\big(F_{L_1} \times L_1(t) + F_{L_2}L_2(t) \times +F_{L_3} \times L_3(t)\big) \tag{10.9}$$

Here $L_1(t), L_2(t), L_3(t)$ are the calculated best fitness values. The AFBER model finds the best solution to secure the transmission of data that helps to gain the trust of the user. After the completion of optimizing the IoT data transmission to the edge computing, the AFBER model performance was analyzed using existing performance metrics.

Experimental Results

The proposed AFBER model performance is employed in the Python software of version 3.10 having a system configuration of processor intel i7, operating system (OS) windows 10, graphic processing unit (GPU) 6 Giga Byte (GB), random access memory (RAM) 16 GB, and memory 1 Tera Byte (TB). The functioning of the planned AFBER model is evaluated in terms throughput and energy consumption for different set of nodes. Table 10.1 presents the functioning of the proposed AFBER method for energy consumption of 0.09, 0.18, 0.26, 0.24, 0.31, 0.30, 0.25, 0.27 for 25, 50, 75, 100, 125, 150, 175, 200 set of nodes. The Tunicate Swarm Algorithm (TSA), Krill Herd Algorithm (KHA), Golden Jacke Optimization (GJO), and traditional Al-Biruni Earth Radius (BER) algorithms were used for the comparisons of proposed AFBER model.

Table 10.2 affords the performance of the proposed AFBER method for throughput for various set of nodes. The model has developed higher throughput compared to the existing algorithms. The TSA, KHA, GJO, and traditional BER algorithm were used for comparisons of the proposed AFBER method and has obtained an improved throughput of 46.52, 23.59, 19.74, 11.07, 8.69, 7.86, 6.76, 5.83 for 25, 50, 75, 100, 125, 150, 175, 200 group of nodes.

Table 10.1 Performance of the proposed AFBER method for energy consumption

No of nodes	Energy consumption				
	TSA	KHA	GJO	BER	Proposed AFBER
25	0.37	0.33	0.24	0.17	0.09
50	0.45	0.38	0.28	0.23	0.18
75	0.47	0.41	0.35	0.30	0.26
100	0.56	0.48	0.36	0.29	0.24
125	0.57	0.50	0.41	0.37	0.31
150	0.60	0.54	0.48	0.35	0.30
175	0.63	0.56	0.46	0.31	0.25
200	0.54	0.49	0.40	0.34	0.27

Table 10.2 Performance of the proposed AFBER method for throughput

No of nodes	Throughput				
	TSA	KHA	GJO	BER	Proposed AFBER
25	35.26	38.21	41.06	43.25	46.52
50	14.58	16.79	19.57	21.04	23.59
75	13.23	14.23	15.58	17.65	19.74
100	7.25	8.42	9.05	10.00	11.07
125	4.67	5.87	6.13	7.21	8.69
150	4.24	5.02	5.47	6.54	7.86
175	3.32	3.86	4.23	5.28	6.76
200	2.87	3.44	3.99	4.21	5.83

Comparative Assessment

The developed AFBER model performance is compared with the existing IB-SEC (Irshad et al. 2023) method for different set of nodes. The comparative assessment provides the details of the proposed AFBER model results with the available method for the secure data transmission from the IoT-based devices to the edge computing. This section explains the improvement of the proposed AFBER model. Table 10.3 indicates the comparative of the proposed AFBER method for energy consumption and has obtained minimal energy consumption of 0.09 J, 0.18 J, 0.26 J, 0.24 J, 0.31 J, 0.30 J, 0.25 J, 0.27 J for 25, 50, 75, 100, 125, 150, 175, 200 set of nodes compared to existing IB-SEC (Irshad et al. 2023) method.

Table 10.4 provides the comparative assessment of the developed AFBER method for throughput metric. The IB-SEC (Irshad et al. 2023) model is used for comparisons with the implemented AFBER method. The performance of the developed AFBER model has given better results for higher throughput of 46.52, 23.59, 19.74, 11.07, 8.69, 7.86, 6.76, 5.83 Kbps for 25, 50, 75, 100, 125, 150, 175, 200 group of nodes.

Table 10.3 Comparative assessment of the proposed AFBER method for energy consumption

No of nodes	Energy consumption (J)	
	IB-SEC (Irshad et al. 2023)	Proposed AFBER
25	0.10334	0.09
50	0.23822	0.18
75	0.31887	0.26
100	0.32665	0.24
125	0.3501	0.31
150	0.333	0.30
175	0.3299	0.25
200	0.31	0.27

Table 10.4 Comparative assessment of the proposed AFBER method for throughput

No of nodes	Throughput (Kbps)	
	IB-SEC (Irshad et al. 2023)	Proposed AFBER
25	44.301	46.52
50	22.434	23.59
75	17.242	19.74
100	9.726	11.07
125	7.23105	8.69
150	6.964	7.86
175	5.76	6.76
200	5.241	5.83

Discussion

The discussion exhibits the drawbacks and problem faced by the existing methodology during the process of securing the cloud computing for IoT-based smart cities along with the benefits of the proposed AFBER method. The existing IB-SEC (Irshad et al. 2023) method failed to handle the multiple attacks in the network that suppressed the security of data transmission. The delay in the user verification and authentication process led path for vulnerable activities. Maintenance of the user trust toward the edge computing is affected by the latency in the data transmission. The traditional algorithms have limited searching ability that affected the performance of transmitting the data. The proposed AFBER method has overcome the problem of existing method and resist to the multiple attacks in the network that improved the stability and security of the data transmission between the edge computing and IoT-based smart cities. The developed AFBER model has obtained as better results having minimal energy consumption of 0.09 J, 0.18 J, 0.26 J, 0.24 J, 0.31 J, 0.30 J, 0.25 J, 0.27 J and for higher throughput of 46.52, 23.59, 19.74, 11.07, 8.69, 7.86, 6.76, 5.83 Kbps for 25, 50, 75, 100, 125, 150, 175, 200 group of nodes compared to the existing IB-SEC (Irshad et al. 2023) model.

Conclusion

The conclusion provides the brief view of the proposed AFBER optimization algorithm used for trust and secure edge computing for IoT data transmission. The active fitness mechanism is incorporated in the BER optimization algorithm to advance the space searching ability of the algorithm that finds the best solution for the secure data transmission. The developed AFBER model has obtained as better results having minimal energy consumption of 0.09 J, 0.18 J, 0.26 J, 0.24 J, 0.31 J, 0.30 J, 0.25 J, 0.27 J and for higher throughput of 46.52, 23.59, 19.74, 11.07, 8.69, 7.86, 6.76, 5.83 Kbps for 25, 50, 75, 100, 125, 150, 175, 200 group of nodes compared to the existing IB-SEC (Irshad et al. 2023) model. Further an advanced optimization technique-based deep learning (DL) algorithm can be incorporated for the prediction of network traffic and vulnerable activities during the process of data transmission that improves the trust and security of the data in edge computing for IoT-based smart cities.

References

Ahmed I, Zhang Y, Jeon G, Lin W, Khosravi MR, Qi L (2022a) A blockchain-and artificial intelligence-enabled smart IoT framework for sustainable city. Int J Intell Syst 37(9):6493–6507

Ahmed A, Abdullah S, Bukhsh M, Ahmad I, Mushtaq Z (2022b) An energy-efficient data aggregation mechanism for IoT secured by blockchain. IEEE Access 10:11404–11419

Ajao LA, Apeh ST (2023) Secure edge computing vulnerabilities in smart cities sustainability using petri net and genetic algorithm-based reinforcement learning. Intell Syst Appl 18:200216

Ali J, Khan MF (2023) A trust-based secure parking allocation for IoT-enabled sustainable smart cities. Sustainability 15(8):6916

Ali B, Hijjawi S, Campbell LH, Gregory MA, Li S (2022) A maturity framework for zero-trust security in multiaccess edge computing. Secur Commun Netw 2022(1):3178760

Huang Y (2024) Smart home system using blockchain technology in green lighting environment in rural areas. Heliyon 10(4):e26620

Irshad RR, Hussain S, Hussain I, Ahmad I, Yousif A, Alwayle IM, Alattab AA, Alalayah KM, Breslin JG, Badr MM, Rodrigues JJ (2023) An intelligent buffalo-based secure edge-enabled computing platform for heterogeneous IoT network in smart cities. IEEE Access 11:69282–69294

Javeed D, Saeed MS, Adil M, Kumar P, Jolfaei A (2024) A federated learning-based zero trust intrusion detection system for Internet of Things. Ad Hoc Netw 162:103540

Khan M, Hatami M, Zhao W, Chen Y (2024) A novel trusted hardware-based scalable security framework for IoT edge devices. Discov Internet Things 4(1):4

Latif R, Ahmed MU, Tahir S, Latif S, Iqbal W, Ahmad A (2022) A novel trust management model for edge computing. Complex Intell Syst 8:3747–3763

Liu Y, Sun Q, Sharma A, Sharma A, Dhiman G (2022) Line monitoring and identification based on roadmap towards edge computing. Wireless Pers Commun 127:441–464

Qayyum T, Trabelsi Z, Waqar Malik A, Hayawi K (2022) Mobility-aware hierarchical fog computing framework for Industrial Internet of Things (IIoT). J Cloud Comput 11(1):72

Saba T, Rehman A, Haseeb K, Alam T, Jeon G (2023) Cloud-edge load balancing distributed protocol for IoE services using swarm intelligence. Clust Comput 26(5):2921–2931

Shu C, Chen Y, Tan C, Luo Y, Dou H (2024) Enhancing trust transfer in supply chain finance: a blockchain-based transitive trust model. J Cloud Comput 13(1):4

Zhou Y, Zhao B, An Y (2022) A novel trusted software base for commercial android devices using secure TF card. Secur Commun Netw 2022(1):6731277

Chapter 11
Gated Deep Reinforcement Learning with Sea Lion Optimization for Detecting Jamming Attacks in Wireless Sensor Networks

S. Prabhu, Hima Bindu Gogineni, Boddepalli Prameela, R. Rana Veer Samara Sihman Bharattej, and D. Navaneetha

Abstract The rapid growth in technology has increased the applications of wireless sensor networks (WSN) in a wide range. The sensor nodes gather the surrounding information and further transmit central storage where the data are secured. However, the security of the data is decreased in the transmission process due to jamming attacks in the network. The existing model tried to generate the jamming attack detection mechanism to clear the disturbances in the data transmitting path, but failed to reach their outcome due to loss of data packets during the transmission. The gated deep reinforcement learning with sea lion optimization (GDRL with SLO) is proposed to minimize the loss of data packets during the transmission process. The gated deep reinforcement learning (GDRL) interpreted the details of the data related to jamming attacks in the network and the sea lion optimization (SLO) model enhanced the global searching ability of the GDRL that exhibited the appropriate jamming attack detection in the WSN. The developed GDRL with SLO has attained

---

S. Prabhu
Department of Electronics and Communication Engineering, Mahendra Institute of Technology, Mallasamudram, India
e-mail: hodece@mahendratech.org

H. B. Gogineni
Department of Information Technology, Anil Neerukonda Institute of Technology and Sciences (ANITS), Visakhapatnam, India

B. Prameela
Department of AI&DS, VIGNAN's Institute of Information and Technology, Visakhapatnam, India

R. R. V. S. S. Bharattej
Doctorate of Business Administration, National Louis University, Tampa, FL, USA

D. Navaneetha (✉)
Associate Professor, Information Technology, Bhoj Reddy Engineering College for Women, Hyderabad, India
e-mail: navaneethareddy22@gmail.com

G. P. Ramesh et al. (eds.), *Empowering IoT with Distributed Edge Computing*, IoT and Wireless Communication Networks, https://doi.org/10.1007/978-981-96-9455-6_11

higher throughput and packet delivery ratio of 94 Mbps and 97% compared to the existing reinforcement learning-based gradient monitored (RLGM) model.

Keywords Data transmission · Energy consumption · Gated deep reinforcement learning · Jamming attack detection · Sea lion optimization algorithm · Wireless sensor network

Introduction

Wireless sensor network (WSN) is form of network where the sensor nodes are circulated in the atmosphere to monitor the physical status of the environment like sound, pressure, temperature or pollution (Zahra et al. 2024). The sensor nodes, communicating protocols, and base station (BS) are the major components which forms the WSN (John et al. 2024). The sensor nodes are responsible for the environment data gathering (Xu et al. 2023). The communicating protocols are responsible for the transmission of data from the sensor nodes to the BS (Almomani et al. 2022). The transmitted data are gathered and stored in the BS and also act as a central point for data receiving and transmitting (Elsadig 2023). The reliability of the WSN depends on the energy management of the sensor nodes (Sivaprakash et al. 2023). The increase in the usage of the WSN has led path for different illegal network activities that affects the security and privacy of the data during the process of transmission (Gebremariam et al. 2023). The disturbance in the data transmission where the data signal get blocked is known as jamming attack (Algarni et al. 2024). The irrelevant signal is transmitted by the attackers where the sensors get confused with the destination of the data transmission that affects the security of the data (Kanagasabapathy et al. 2022). The contribution of the proposed GDRL with SLO algorithm for the detection of jamming attack in WSN is given below as follows:

- The research focused on improving the jamming attack detection mechanism for the safe and secure WNS data transmission without any interruption.
- The gated deep reinforcement learning (GDRL) algorithm is used for interpreting the details of the WSN based on its sequential decision-making capability for network jamming attack detection.
- The sea lion optimization (SLO) algorithm advanced the searching ability of the GDRL model by optimizing its parameters that increased the performance of detecting the jamming attack in WSN.

The details information of the benefits and limitations of the existing methodologies used for the detection of jamming attacks in WSN are discussed as follows: Lyu et al. (2022) presented an advanced optimization model for the localization of the jamming attacks in WSN. The Beetle Antenna Search (BAS) algorithm was used for optimizing the security of the WSN by incorporating the adaptive step size strategy. The model located the cause for the jamming in the network based on dynamic position update mechanism of the jamming that enhanced the performance of network jam detection model. However, the latency in the transmission of data led to vulnerable activities in the network that minimized the security of the WSN data. Arivunambi and Paramarthalingam (2022) developed an intelligent smile mold (ISM) model for the detection of WSN jamming attacks. The model identified the range between the source and destination nodes for the detection of optimal path for the transmission of data. The jamming attacks in the network get normalized using the ISM algorithm that advanced the network detection ability of the model in WSN. However, the model took huge energy during the transmission process that minimized the lifetime of the network and led path for vulnerable activities. Meleshko and Desnitsky (2024) implemented a network attacks detection model for WSN based on the self-organized decentralization technique. The model interpretated the details of the WSN to detect the jamming attacks present in the network. The self-organized decentralize method predicted the occurrence of jamming attacks in the network based on the previous data history that improved the security of the data transmission in the WSN. However, the model performance was affected due to loss of data packets during the transmission that caused heavy jamming attack in the network.

Gowdhaman and Dhanapal (2022) presented a deep neural network (DNN) for the detection of attacks in the WSN. The jamming attack data were used for the detection of WSN attacks. The data were preprocessed by normalization technique and then significant information was selected from the data and then the DNN model interpretated the details of selected feature for the categorization of data into different class and detected the appropriate jamming attack in the WSN. However, the model failed to handle the details of the verification process that suppressed the performance of the model in detecting the jamming attacks in the network. Saleh et al. (2024) developed a WSN jamming attack detection system using stochastic gradient descent (SGD) model. The principal component analysis (PCA) was used for handling the feature dimensionality that advanced the interpretability of the model. The SGD enhanced the jamming attack detection performance with the incorporation of Gaussian Naïve Bayes (GNB) that increased the throughput of the WSN. However, the model struggled to adapt for the sudden change in the data transmission that led way for the jamming attack in the network. Ghelani et al. (2024) presented a reinforcement learning-based gradient monitored (RLGM) model for detecting jamming attack in the network. The model handled the gradient variance problem during the training process. The irregular interruption in the detection system was suppressed using RLGM model that enhanced the total performance of the jamming attack detection mechanism. However, the unauthorized access in the network affected the security of the data transmission and minimized the stability of jamming attack detection system. The existing techniques faced various hardships during the process of detecting the jamming attack in the WSN. The delay in the process of transmitting the data led to packet loss during the transmission of data.

The rest of the sections present in this research paper is formatted as shown below: Section "Proposed Methodology" provides the proposed methodology. Section "Experimental Results" exhibits the experimental results. Section "Discussion" consists of discussion, and Section "Conclusion" gives the discussion.

Proposed Methodology

The GDRL with SLO model is proposed for the jamming attack detection in the WSN. The parameters and weights of the nodes are initialized for the network. The model gets trained with by splitting the data and further the GDRL with SLO model upgraded the weights of the nodes to identify the optimal explanation for the jamming attack. Figure 11.1 provides the flowchart of the proposed GDRL with SLO for jamming attack detection.

The flow of the jamming attack detection in WSN process is explained here:

- Start: The jamming attack detection process begins.
- Create the network: The interaction between the system and network is activated with the sensor nodes.
- Initialization of parameters and assign global weight to all the nodes in the network: Here the network parameters get initialized where the global weights get scattered in the existing nodes.
- Train a local model by splitting the data: The data get separated to get trained with each and every portion is used for training every single node.
- Calculate the new weight values using GDRL with SLO and detect the jammer: The proposed GDRL-based SLO provides the upgraded weights of each parameter values based on the data that identifies and detects the jammer node in the network.
- The path with the jammer node is eliminated: After detecting the jammer node, the jamming path gets eliminated to secure the data.
- End: The jamming attack detection process stops.

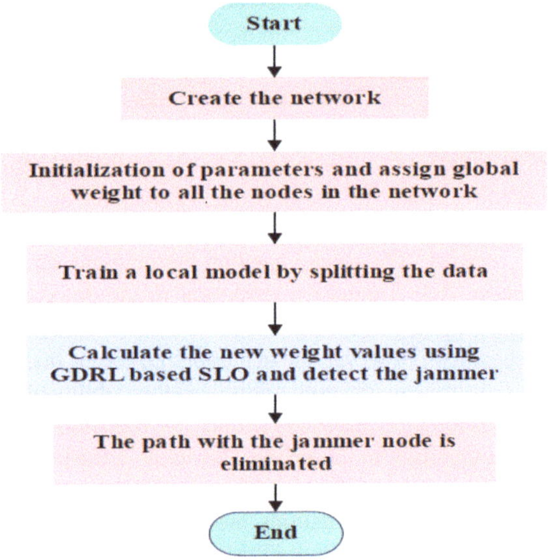

Fig. 11.1 Flowchart of the proposed GDRL with SLO for jamming attack detection

GDRL-Based SLO for the Detection of Jamming Attack in WSN

The GDRL is the algorithm used for the detection of jamming attack in the WSN based on nodes interaction. The model provides the sequential decision-making ability for the detection of jamming attack. The GDRL model is trained with the data of the sensor nodes and indicates the data representation in the vector form. The SLO is incorporated with the GDRL to increase the training of the model by optimizing the parameters that helps for the upgradation of the weights. The sea lion prey detection strategy, hunting communication (vocalization), and attacking mechanisms were used by the SLO algorithm to upgrade the weights of the parameters. The details of the three stages of the SLO are given below:

Detection and Tracking Stage. The sea lion utilizes their whiskers to locate their prey based on the shape, position and size of the objects that are closer. The whiskers are positioned against the flow of water, the sea lion detect their prey location based on the vibration difference. The sea lion is taken as a leader and then other member of their troop restructure their location to target the prey. The mathematical representation of the detecting the target prey is given using Eq. (11.1),

$$\overrightarrow{\text{Dist}} = \left|\overrightarrow{2B} \cdot \overrightarrow{P(t)} - \overrightarrow{\text{SL}(t)}\right| \tag{11.1}$$

Here $\overrightarrow{\text{SL}(t)}$ gives the position vector of sea lion, $\overrightarrow{P(t)}$ represents the target prey, $\overrightarrow{\text{Dist}}$ exhibits the gap between the target and sea lion, B gives the random vector and t shows the present iteration. The sea lion gets closer to their prey using the next hunting strategy using Eq. (11.2),

$$\overrightarrow{\text{SL}(t+1)} = \overrightarrow{P(t)} - \overrightarrow{\text{Dist}} \cdot \vec{C} \tag{11.2}$$

The main leader led the path in the direction of the prey by SLO algorithm and surround the target prey within the range.

Vocalization Stage. The sea lion lives both on water and land, where the vocalization travels four times speed in water compared to air. Based on the vocalization range the sea horse communicate with their hunting troops. The members of sea horse preset in the shores are invited based on their voice range. After capturing the prey, it is taken to the shore that is closer. The mathematical equation used for encircling the prey is given using Eqs. (11.3)–(11.5),

$$\overrightarrow{\text{SP}_{\text{leader}}} = \left|\left(\vec{V_1}\left(1 + \vec{V_2}\right) / \vec{V_2}\right)\right| \tag{11.3}$$

$$\vec{V_1} = \sin\theta \tag{11.4}$$

$$\overrightarrow{V_2} = \sin \emptyset \tag{11.5}$$

Here $\overrightarrow{V_2}$ gives the speed of sound in air, $\overrightarrow{V_1}$ provides the speed of sound in water, $\sin\theta$ indicates the air-to-air interaction and $\sin\emptyset$ shows the interaction between air and water.

Exploitation Stage. The sea lion discovers and encircle their individual prey and gives signal to the remaining members of the troop to get the location. The discovered prey is the optimal explanation available in the current situation. Further the upgraded location of the prey is obtained using Eq. (11.6),

$$\overrightarrow{\mathrm{SL}(t+1)} = \left|\overrightarrow{P(t)} - \overrightarrow{\mathrm{SL}(t)}\right| \cdot \cos(2\pi m) + \overrightarrow{P(t)} \tag{11.6}$$

Here m provides the random value.

Exploration Stage. The sea lion utilizes their zigzag swimming and whisker tactics for searching the prey randomly. The localization is modified based on best searching performance of the sea lion. The searchers alter their locations using best sea lion value. The global solution is obtained using Eqs. (11.7) and (11.8),

$$\overrightarrow{\mathrm{Dlst}} = \left|\overrightarrow{2B} \cdot \overrightarrow{\mathrm{SL}}_{\mathrm{rnd}}(t) - \overrightarrow{\mathrm{SL}(t)}\right| \tag{11.7}$$

$$\overrightarrow{\mathrm{SL}(t+1)} = \overrightarrow{\mathrm{SL}}_{\mathrm{rnd}}(t) - \overrightarrow{\mathrm{Dlst}} \cdot \vec{C} \tag{11.8}$$

Here $\overrightarrow{\mathrm{SL}}_{\mathrm{rnd}}(t)$ provides the sea lion randomly choose from the existing troop. The SLO identified the best explanation for the upgradation of the searcher position. The jamming attack is detected using GDRL with SLO algorithm.

Experimental Results

The developed GDRL with SLO model functioning is displayed in the Python software of version 3.9 having a system configuration of processor intel i7, random access memory (RAM) 16 giga byte (GB), operating system (OS) windows 10, graphic processing unit (GPU) 6 GB, and memory 1 Tera Byte (TB). Table 11.1 exhibits the performance of the proposed GDRL with SLO method for delay (s) metric for various set of nodes. The associative reinforcement learning (ARL), deep Q-learning (DQL), federated reinforcement learning (FRL), and traditional deep reinforcement learning (DRL) are used to comparison. The developed GDRL with SLO model has obtained minimal delay of 0.04, 0.07, 0.08, 0.10, 0.13, 0.18, 0.19, 0.20, 0.23, and 1.42 s for 10, 20, 30, 40, 50, 60, 70, 80, 90, and 100 number of nodes.

Table 11.1 Performance of the developed GDRL with SLO method for delay (s)

Methodology	Number of nodes									
	10	20	30	40	50	60	70	80	90	100
ARL	0.13	0.18	0.22	0.25	0.27	0.32	0.36	0.39	0.41	2.87
DQL	0.10	0.14	0.19	0.18	0.24	0.29	0.33	0.34	0.35	2.46
FRL	0.08	0.11	0.15	0.16	0.19	0.25	0.27	0.28	0.30	2.03
DRL	0.06	0.09	0.12	0.13	0.15	0.21	0.23	0.25	0.26	1.89
Proposed GDRL with SLO	0.04	0.07	0.08	0.10	0.13	0.18	0.19	0.20	0.23	1.42

Table 11.2 explains the performance of the proposed GDRL with SLO method for energy consumption (J). The developed GDRL with SLO model has obtained less energy consumption of 85, 100, 124, 133,158, 182, 215, 244, 291, and 286 J for 10, 20, 30, 40, 50, 60, 70, 80, 90, and 100 number of nodes compared to the existing ARL, DQL, FRL, and traditional DRL methods.

Table 11.3 indicates the performance of the proposed GDRL with SLO method for packet delivery ratio (%). The ARL, DQL, FRL and traditional DRL were used for the comparison of the GDRL with SLO. The developed GDRL with SLO model has exhibited higher packet delivery ratio of 77, 84, 86, 90, 93, 94, 92, 96, 97, and 96% for 10, 20, 30, 40, 50, 60, 70, 80, 90, and 100 number of nodes.

Table 11.4 gives the performance of the proposed GDRL with SLO method for throughput (Mbps) for different group of nodes. The GDRL with SLO model has

Table 11.2 Performance of the proposed GDRL with SLO method for energy consumption (J)

Methodology	Number of nodes									
	10	20	30	40	50	60	70	80	90	100
ARL	105	129	142	159	181	213	248	270	348	329
DQL	97	125	139	151	175	204	236	263	334	316
FRL	93	117	134	145	169	196	227	254	321	307
DRL	89	106	128	137	162	189	220	249	298	295
Proposed GDRL with SLO	85	100	124	133	158	182	215	244	291	286

Table 11.3 Performance of the developed GDRL with SLO model for packet delivery ratio (%)

Methodology	Number of nodes									
	10	20	30	40	50	60	70	80	90	100
ARL	58	64	68	71	79	81	77	79	81	76
DQL	61	69	72	75	81	84	80	83	85	80
FRL	65	73	78	79	85	88	85	89	91	87
DRL	71	79	82	85	89	90	89	91	94	93
Proposed GDRL with SLO	77	84	86	90	93	94	92	96	97	96

Table 11.4 Performance of the proposed GDRL with SLO method for throughput (Mbps)

Methodology	Number of nodes									
	10	20	30	40	50	60	70	80	90	100
ARL	54	58	64	69	71	75	79	81	82	81
DQL	59	61	68	74	77	79	82	85	86	85
FRL	64	67	73	79	83	85	86	88	89	88
DRL	69	71	78	83	88	89	89	90	91	91
Proposed GDRL with SLO	74	79	82	87	93	92	91	94	95	94

obtained a greater throughput of 74, 79, 82, 87, 93, 92, 91, 94, 95, and 94 Mbps for 10, 20, 30, 40, 50, 60, 70, 80, 90, and 100 number of nodes compared to the existing ARL, DQL, FRL, and traditional DRL algorithms.

Comparative Assessment

The functioning of the proposed GDRL with SLO algorithm is assessed with the existing RLGM (Ghelani et al. 2024) technique. The advancement in the searching ability has given appropriate detection of jamming attack in the WSN that protected the data. Table 11.5 exhibits the comparative assessment of the proposed GDRL with SLO method with the RLGM (Ghelani et al. 2024) model for various number of nodes.

Discussion

This section provides the details of the hardships faced the existing RLGM (Ghelani et al. 2024) method during the detection of jamming attack in the WSN along with the benefits of the developed GDRL with SLO method. The existing algorithm failed to reduce the delay in the data transmission due to vulnerable activities that reduced the security of WSN data. The unauthorized access in the network affected the security of the data transmission and minimized the stability of jamming attack detection system. The proposed GDRL with SLO overcome the existing issues by finding the best solution for the detecting the jamming attack in the network. The developed GDRL with SLO has obtained increased throughput of 74, 79, 82, 87, 93, 92, 91, 94, 95, and 94 Mbps. Minimal delay of 0.04, 0.07, 0.08, 0.10, 0.13, 0.18, 0.19, 0.20, 0.23, and 1.42 s. Less energy consumption of 85, 100, 124, 133, 158, 182, 215, 244, 291, and 286 J. Higher packet delivery ratio of 77, 84, 86, 90, 93, 94, 92, 96, 97, and 96% for 10, 20, 30, 40, 50, 60, 70, 80, 90, and 100 number of nodes.

Table 11.5 Comparative assessment of the proposed GDRL with SLO method with RLGM (Ghelani et al. 2024)

No of nodes	Delay (s)		Energy consumption (J)		Packet delivery ratio (%)		Throughput (Mbps)	
	RLGM (Ghelani et al. 2024)	Proposed GDRL with SLO	RLGM (Ghelani et al. 2024)	Proposed GDRL with SLO	RLGM (Ghelani et al. 2024)	Proposed GDRL with SLO	RLGM (Ghelani et al. 2024)	Proposed GDRL with SLO
10	0.05	0.04	90	85	74	77	72	74
20	0.08	0.07	105	100	79	84	77	79
30	0.09	0.08	120	124	82	86	79	82
40	0.12	0.10	139	133	87	90	85	87
50	0.15	0.13	160	158	90	93	89	93
60	0.2	0.18	190	182	92	94	90	92
70	0.22	0.19	220	215	91	92	90	91
80	0.24	0.20	250	244	94	96	92	94
90	0.25	0.23	298	291	94	97	92	95
100	1.68	1.42	293	286	95	96	93	94

Conclusion

The overall research concludes that the GDRL with SLO algorithm is proposed for the detection of jamming attack in the WSN. The developed GDRL with SLO model optimized the parameters of the network for the detection of appropriate jamming attack in WSN during data transmission. The GDLR model was used for jamming attack detection based on the features of the network. The SLO algorithm advanced the searching ability of the GDLR model by optimizing the parameters. The developed GDRL with SLO model has obtained higher throughput and packet delivery ratio of 94 Mbps and 97% for 100 number of nodes. The proposed GDRL with SLO algorithm has exhibited lower energy consumption and delay of 286 J and 0.23 s for 100 number of nodes compared to the existing RLGM (Ghelani et al. 2024) model. Further, the performance of the WNS jamming attack detection model be advanced using various modified version of deep learning model-based optimization techniques to detect the appropriate jamming node for the smooth and safe WNS data transmission.

References

Algarni A, Acarer T, Ahmad Z (2024) An edge computing-based preventive framework with machine learning-integration for anomaly detection and risk management in maritime wireless communications. IEEE Access 12:53646–53663

Almomani I, Ahmed M, Kosmanos D, Alkhayer A, Maglaras L (2022) An efficient localization and avoidance method of jammers in vehicular ad hoc networks. IEEE Access 10:131640–131655

Arivunambi A, Paramarthalingam A (2022) Intelligent slime mold algorithm for proficient jamming attack detection in wireless sensor network. Glob Transit Proc 3(2):386–391

Elsadig MA (2023) Detection of denial-of-service attack in wireless sensor networks: a lightweight machine learning approach. IEEE Access 11:83537–83552

Gebremariam GG, Panda J, Indu S (2023) Design of advanced intrusion detection systems based on hybrid machine learning techniques in hierarchically wireless sensor networks. Connect Sci 35(1):2246703

Ghelani J, Gharia P, El-Ocla H (2024) Gradient monitored reinforcement learning for jamming attack detection in FANETs. IEEE Access 12:23081–23095

Gowdhaman V, Dhanapal R (2022) An intrusion detection system for wireless sensor networks using deep neural network. Soft Comput 26(23):13059–13067

John A, Isnin IFB, Madni SHH, Faheem M (2024) Cluster-based wireless sensor network framework for denial-of-service attack detection based on variable selection ensemble machine learning algorithms. Intell Syst Appl 22:200381

Kanagasabapathy PMK, Poornachary KV, Murugan S, Natesan A, Ponnusamy V (2022) Rapid jamming detection approach based on fuzzy in WSN. Int J Commun Syst 35(2):e4205

Lyu Y, Mo Y, Yue S, Liu W (2022) Improved beetle antennae algorithm based on localization for jamming attack in wireless sensor networks. IEEE Access 10:13071–13088

Meleshko A, Desnitsky V (2024) The modeling and detection of attacks in role-based self-organized decentralized wireless sensor networks. Telecom 5(1):145–175

Saleh HM, Marouane H, Fakhfakh A (2024) Stochastic gradient descent intrusions detection for wireless sensor network attack detection system using machine learning. IEEE Access 12:3825–3836

Sivaprakash S, Anbazhagu UV, Perumal I, Kumar VV, Mahesh TR, Guluwadi S (2023) Analysis and attack detection in GSM mobile network with an intelligent jammer using ANFIS classifier. IEEE Access 11:118962–118972

Xu B, Lu M, Zhang H (2023) Multi-agent modeling and jamming-aware routing protocols for movable-jammer-affected WSNs. Sensors 23(8):3846

Zahra FT, Bostanci YS, Soyturk M (2024) LSTM-based jamming detection and forecasting model using transport and application layer parameters in Wi-Fi based IoT systems. IEEE Access 12:32944–32958

Chapter 12
Roulette Wheel-Based Multiverse Optimization Algorithm with Opposition-Based Learning for Multidata Collection Task in Wireless Sensor Networks for Smart Agriculture

P. Jaya Prakash, K. Naresh, R. Raja Kumar, G. Tagore Sai Prasad, and B. Ramakantha Reddy

Abstract The advancement in technology has created smart cities, smart gadgets, and other applications where the data get collected form the wireless sensor networks (WSN) for the analysis of requirements of the users. The sensor nodes are user for collecting the soil, water, and temperature data of the agriculture filed for the automatic analysis of agriculture land condition. The existing methods tried to automate the analysis of the agricultural filed to analyze the crop productivity but failed to reach the expected outcome due to energy drain in the sensor nodes. To solve the existing problem the Roulette wheel-based multiverse optimization algorithm with opposition-based learning (RWMVO with OBL) technique is used for the selection of fittest sensor nodes from the available resources that increased the life span of the sensors for multidata collection task for smart agriculture. The OBL algorithm was incorporated for advancing the learning ability of RWMVO. The performance of the developed RWMVO with OBL has obtained higher throughput of 5.68 Kbps, lower energy utilization of 9.05 mW, and less time delay of 795 ms for 500 rounds compared to the existing energy aware software-defined network (EASDN) model.

P. Jaya Prakash (✉) · K. Naresh · R. Raja Kumar
Department of Information Technology, Sri Venkateswara College of Engineering, Tirupati, India
e-mail: pokalajayaprakash@gmail.com

R. Raja Kumar
e-mail: rajakumar.r@svcolleges.edu.in

G. Tagore Sai Prasad
Department of Computer Science and Engineering, Koneru Lakshmaiah Education Foundation, Guntur, India

B. Ramakantha Reddy
Department of Computer Science and Engineering (AIML), Sri Venkateswara College of Engineering, Tirupati, India

G. P. Ramesh et al. (eds.), *Empowering IoT with Distributed Edge Computing*, IoT and Wireless Communication Networks, https://doi.org/10.1007/978-981-96-9455-6_12

Keywords Edge computing · Energy utilization · Smart agriculture · Multidata collection task · Roulette wheel-based multiverse optimization algorithm with opposition-based learning · Wireless sensor network

Introduction

The wireless sensor network (WSN) refers to the distributed sensor nodes in the environment that collects the surrounding information and transfer them to the edge computing (Sudhakar and Anne 2024). The WNS is having wide range of application in various fields such as military for monitoring the movements of enemies, health care for monitoring the patients, industries for the smart factories and agriculture for analyzing the soil moisture with crop monitoring (Chamara et al. 2023). The sensor node is the core unit of the WSN which is capable of handling the physical atmosphere data. The nodes comprise the batteries where the energy get drained due to limited energy capacity. The sensor nodes transfer the collected and aggregated data information to the base station (BS) (Wen et al. 2023; Rosero-Montalvo et al. 2023). Then the data are further sent to the edge computing where the data are stored and secured. The edge computing and WSN are essential for the multidata collection in smart agriculture for data processing and monitoring (Tseng et al. 2023). The agricultural lands are surrounded by the sensor nodes that monitors various factors of the surroundings that helps for the management of agricultural fields (Akhter and Sofi 2022). The edge computing and WSN are mutually depending on each other for the collecting the multidata from the agricultural filed (Križanović et al. 2023; Ali et al. 2023). The optimization of the sensor node performance helps to reduce the consumption of energy during the process of multidata collection (Makondo et al. 2024). The estimation of sensor nodes fitness provides the best sensor nodes for the available resources. The contribution of the proposed RWMVO with OBL algorithm for multidata collection task in WSN for smart agriculture is given below as follows:

- The research aims to advance the multidata collection task of the WSN to analyze the condition of the agriculture filed that helps to increase the agriculture productivity.
- The Roulette Wheel-based Multiverse Optimization Algorithm (RWMVO) was used for the selection of optimal sensor nodes for the available resources for multidata collection task in WSN for smart agriculture.
- The Opposition-Based Learning (OBL) technique was incorporated in the RWMVO model that advanced the searching ability of the model to find the most suitable sensor nodes based on the fitness value that retains the sensor node energy during the data collection process.
- The Roulette Wheel-based Multiverse Optimization Algorithm with Opposition-Based Learning (RWMVO with OBL) technique has identified the fittest sensor nodes for the multidata collection with minimal energy utilization for the smart agriculture.

Literature Survey

This section exhibits the benefits and problems faced by the existing models during multitask data collection based on WSN and edge computing for smart agriculture. Raghuvanshi et al. (2022) implemented a machine learning (ML)-based smart irrigation system using WSN data. The data were normalized to obtain standard range for value of the data and then principal component analysis (PCA) minimized the dimensionalities of the data features. The support vector machine (SVM) algorithm secured the irrigation WNS data from vulnerable access. However, the models failed to handle the multiple attacks in the network that led to delay in the transmission of data and reduced the performance of smart irrigation monitoring system. Aldossary et al. (2024) developed an efficient smart agriculture monitoring model using IoT data. The long short-term memory (LSTM)-based Inception algorithm was used for the classifying the agricultural data. The model determined the soil quality based on the details of data that advanced the efficiency and performance of smart agricultural monitoring system. However, the performance of the model was affected due to latency in the process of collecting the details from the nodes that consumed huge energy and suppressed the quality of smart agriculture monitoring system. Bindu et al. (2023) presented a graph theory based smart agriculture management system using Clustered WSN (CWSN). The irrigation water system was monitored by the model to identify the water flow with the feasible senor node energy utilization. The graph theory handled the transiting data that enhanced the performance of the smart agriculture management system. However, the performance of the model was affected by the unauthorized access in the network that suppressed the stability and security of smart agriculture mechanism.

Itoo et al. (2023) implemented a smart agriculture monitoring system based on lightweighted authentication. The privacy preserving mechanism was for the secure transmission of collected data by the sensors. The elliptic curve cryptography (ECC) was the technique used for securing the data. The presented mutual authentication process had advanced the performance of the smart agriculture management system. However, the stability of the model was affected due to higher energy consumption by the sensor nodes that minimized the performance of the smart agriculture management system. Saba et al. (2023) developed a block chain-based system using WNS for smart agriculture monitoring. The incorporation of block chain collected the agricultural data and secure it by providing the authentication. The fitness function was used for optimizing the energy consumption of the sensor nodes during the collection of data that advanced the monitoring ability of the model for smart agricultural system. However, the model performance gets affected due to vulnerable activities in the network that suppressed the data gathering performance in smart agricultural system. Ahmed (2023) presented an energy aware model-based SN for the smart agriculture management. The energy aware software defined network (EASDN) model was used for the selecting the fittest cluster head from the sensor nodes that save the energy during the data collecting process and improved the performance of smart agriculture management system. However, the model performance was affected due to traffic in the data transmission path that suppressed the stability of the smart agriculture management model. The existing models faced various difficulties during the process of multidata collection for smart agriculture. The model performance gets affected due to vulnerable activities in the network that suppressed the data gathering performance in smart agricultural system. The models failed to handle the multiple attacks in the network that led to delay in the transmission of data and reduced the performance of WSN during the data collection. The stability of the model was affected due to higher energy consumption by the sensor nodes that minimized the performance of multidata collecting task.

Proposed Methodology

The RWMVO with OBL algorithm is proposed for the edge computing based multidata collection task using WSN for smart agriculture. The edge server and WSN platform are mutually dependent on each other for data collection. The multiple task that needs to be executed are analyzed to know the interactions of the sensor node's ability for the assigned task. Based on the previous data history of data collects the current task are assigned to the sensor nodes with the available resources. Figure 12.1 provides the frame work of the proposed RWMVO with OBL methodology.

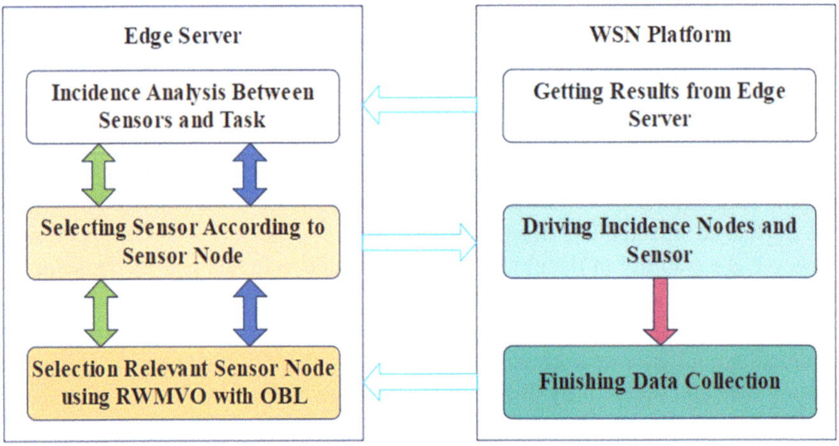

Fig. 12.1 Frame work of the proposed RWMVO with OBL methodology

The Flow of the Edge Server and WSN Platform Interaction for Multidata Collection Task is Given Below as Follows

Edge Server. The edge server is playing important role for the collection and aggregation of data before transmitting to the database. The continuous data are generated with huge quantity like smart agriculture for the analysis purpose. The filtering and cleaning of data before transmitting is processed in the edge server.

Incidence analysis between sensor and task. The relationship and interaction of the various sensor nodes and assigning the task that need to be executed. This stage provides the details of proper requirements for the tasks.

Selecting sensor according to sensor nodes. The appropriate sensor is selected by the edge server after incidence analysis. The related specified sensor is activated in this stage by optimizing the available resources.

Selection of relevant sensor nodes using RWMVO with OBL. The most suitable sensor nodes are elected using RWMVO with OBL algorithm. In this stage, the RWMVO with OBL algorithm finds the most suitable solution for improving energy efficiency of the sensor nodes.

WSN Platform. The WSN are deployed for the collection of surrounding multidata for smart agriculture. The wide range of sensor nodes are used for gathering the multidata and transmitting them for the process of analysis.

Getting results from edge server. The information present in the edge servers is transferred to the WSN platform. The data contain the details of selected sensor nodes on the basis of their previous history of optimization and the analysis. The maintenance of energy of the sensor nodes is the major task in the WSN to prolong the function of the sensors.

Driving incidence nodes and sensor. Based on the details of received data, the sensor nodes get activated by the WSN platform. The initialization for the multi-task data collection takes place in this stage by the selected sensor nodes.

Finishing data collection. After completion of multitask data collection, the WNS platform collects the data from the activated sensor nodes. The gathered data are processed and then transmitted to the analysis purpose.

Proposed RWMVO with OBL for the Selecting the Fittest Sensor Nodes

The selection of the suitable sensor nodes plays a major role for the multidata collection task for smart agriculture. The MVO with OBL algorithm is proposed for the selecting the best sensor nodes base on their fitness values within the available resources. The MVO works on the process of object transition across various universe based on the black or white hole tunnel. Here the tunnels are involving two various universe namely black hole that contains the universe with lower inflation and white hole that have universe with higher inflation. The model enables the migration of the object from one universe to another universe efficiently. The solution matrix of MVO is given in Eq. (12.1),

$$U = \begin{bmatrix} x_1^1 & x_1^2 & \ldots & x_1^d \\ x_2^1 & x_2^2 & \ldots & x_2^d \\ \ldots & \ldots & \ldots & \ldots \\ x_n^1 & x_n^2 & \ldots & x_n^d \end{bmatrix} \tag{12.1}$$

The universes get ranked based on the rates of inflation during each iteration. The roulette wheel technique is incorporated for the selection of universe which contains white holes. The dynamics of the black and white hole tunnel is described and outcomes transfer the objects in-between the universe is seized using mathematical expression Eq. (12.2),

$$x_i^j = \begin{cases} x_k^j r_1 < \mathrm{NI}(U_i), \\ x_i^j r_1 \geq \mathrm{NI}(U_i) \end{cases} \tag{12.2}$$

Here, x_i^j exhibits the importance of the jth parameter with ith universe. $\mathrm{NI}(U_i)$ gives the rate of normalized inflation of universe U_i. x_k^j represents the jth parameter with kth universe for the selected roulette wheel technique. The diversity in the universe is maintained by advancing the exploitation phase by incorporating the wormholes within every universe for randomly transported objects across the spatial dimension. The white holes indicate the transmission of objects from one universe to another through wormholes that randomly changes the object composition within the universe. The mathematical representation of the randomly change object composition is given using Eq. (12.3),

$$x_i^j = \begin{cases} x_j + \mathrm{TDR} \times \left(\left(\mathrm{ub}_j - \mathrm{lb}_j\right) \times r_4 + \mathrm{lb}_j\right) & r_3 < 0.5, \\ x_j - \mathrm{TDR} \times \left(\left(\mathrm{ub}_j - \mathrm{lb}_j\right) \times r_4 + \mathrm{lb}_j\right) & r_3 \geq 0.5 \\ x_i^j r_2 \geq \mathrm{WEP} \end{cases} r_2 < \mathrm{WEP}, \tag{12.3}$$

Here, x_j gives the best solution for jth parameter universe. ub_j and lb_j provide the upper and lower boundary. r_2, r_3, and r_4 indicate the randomly selected values and coefficient is given by traveling distance rate (TDR) and WEP gives the wormhole existence probability. The adaptive nature of the coefficients is prepared to attain equilibrium between exploration and exploitation stage of the MVO. The coefficients are obtained using Eqs. (12.4) and (12.5),

$$\mathrm{TDR} = 1 - \frac{l^{1/p}}{L^{1/p'}} \tag{12.4}$$

$$\mathrm{WEP} = \min + l \times \left(\frac{\max - \min}{L}\right) \tag{12.5}$$

Here, p exhibits the exploitation value for the iterations, l gives the present iteration value, L shows the maximum iteration value. The solution for the selecting the optimal sensor node is obtained by the random initialization of the RWMVO with OBL technique on the basis of fitness value. The highly performing sensor nodes are retained and the lower performance sensor nodes get eliminated. The explanation of the OBL is obtained using Eq. (12.6),

The RWMVO performance is upgraded using OBL technique by generating the new optimal solution for selecting the best sensor nodes using Eqs. (12.7) and (12.8),

$$x_j^{\mathrm{OBL}} = \mathrm{ub}_j - \mathrm{lb}_j - x_j \tag{12.6}$$

$$U_i = \begin{cases} U_i^{\mathrm{OBL}} & r_5 \le C_{\mathrm{OBL}}, \\ U_i^{\mathrm{MVO}} & r_5 > C_{\mathrm{OBL}} \end{cases} \tag{12.7}$$

$$C_{\mathrm{OBL}} = -\left(\frac{1}{L}\right)^2 + 2\left(\frac{1}{L}\right) \tag{12.8}$$

Here, U_i^{OBL} indicates the ith solution obtained by the OBL technique, U_i^{MVO} shows the ith solution provided by the MVO algorithm. The selected optimal sensor nodes collect the data from the agriculture land and transmit to the edge computing. The performance of multidata collection task is improved by using RWMVO with OBL model for smart agriculture.

Experimental Results

The performance of the proposed RWMVO with OBL model is implemented in the MATLAB software of version R2020b and the configuration of system contains the processor intel core i7, Operating System (OS) Windows 10, Random Access Memory (RAM) 16 Giga Byte (GB), Graphic Processing Unit (GPU) 6 GB, respectively. The throughput, energy utilization, and time delay are the metrics used for the evaluation of the proposed model. Table 12.1 exhibits the performance of the proposed RWMVO with OBL method for throughput and energy consumption. Global Neighborhood Algorithm (GNA), Ringed Seal Search (RSS), Shuffled Frog Leaping Algorithm (SFLA), and traditional Multiverse Optimization (MVO) are compared with proposed RWMVO with OBL algorithm. The developed RWMVO

Table 12.1 Performance of the proposed RWMVO with OBL method for throughput and energy consumption

Methodology	Number of rounds = 500	
	Throughput (Kbps)	Energy utilization (mW)
GNA	3.37	12.53
RSS	3.95	11.98
SFLA	4.23	11.21
MVO	5.04	10.24
Proposed RWMVO with OBL	5.68	9.05

Table 12.2 Performance of the proposed RWMVO with OBL method for time delay (ms)

Methodology	Number of rounds			
	500	1000	1500	2000
GNA	1164	1037	895	866
RSS	979	924	786	758
SFLA	896	862	701	697
MVO	847	803	741	654
Proposed RWMVO with OBL	795	701	683	599

with OBL algorithm has obtained higher throughput of 5.68 Kilo bites per second (Kbps) and less energy utilization of 9.05 milli Watt (mW) for 500 number of rounds.

Table 12.2 illustrates the performance of the proposed RWMVO with OBL method for time delay. The developed RWMVO with OBL has obtained lower time delay of 795milli seconds (ms) for 500 rounds, 701 ms for 1000 rounds, 683 ms for 1500 rounds, and 599 ms for 2000 rounds. The RWMVO with OBL algorithm has obtained better results for time delay compared to existing GNA, RSS, SFLA, and traditional MVO.

Comparative Assessment

The proposed RWMVO with OBL model performance is compared and assessed with the existing EASDN (Ahmed 2023) algorithm. This section provides the information of the advancement in the proposed RWMVO with OBL model compared to the existing methods for the multdata collection task in WSN for smart agriculture. Table 12.3 provides the comparative assessment for throughput and energy utilization. The proposed RWMVO with OBL algorithm has obtained higher throughput of 5.68 Kbps and lower energy consumption of 9.05 mW compared to the existing EDSDN (Ahmed 2023) model for 500 number of rounds.

Table 12.3 Comparative assessment for throughput and energy utilization

Methodology	Number of rounds = 500	
	Throughput (Kbps)	Energy Utilization (mW)
EASDN (Ahmed 2023)	4.01	11.73
Proposed RWMVO with OBL	5.68	9.05

Table 12.4 Comparative assessment for time delay (ms)

Methodology	Number of rounds			
	500	1000	1500	2000
EASDN (Ahmed 2023)	824	774	738	651
Proposed RWMVO with OBL	795	701	683	599

Table 12.4 illustrates the comparative assessment of the proposed RWMVO with OBL algorithm for time delay metric. The developed RWMVO with OBL algorithm has exhibited the minimal time delay of 795 ms for 500 rounds, 701 ms for 1000 rounds, 683 ms for 1500 rounds, and 599 ms for 2000 rounds compared to the existing EASDN (Ahmed 2023).

Discussion

The discussion section provides the detailed information of the problems faced the existing models during the process of multidata collection task for smart agriculture along with the advantages of the proposed RWMVO with OBL algorithm. The EASDN (Ahmed 2023) model failed to handle the irrelevant access in the network that affected the multi-data collection due to loss of energy in the sensor node causing data packet loss and affected the performance of WSN. The sensor energy drained suddenly due to lack of appropriate optimal solution for multidata collection task. The proposed RWMVO with OBL algorithm has overcome the problem sensor node selection that retained the sensor node energy during the process of multidata collection. The developed RWMVO with OBL algorithm has exhibited the minimal time delay of 795 ms for 500 rounds, 701 ms for 1000 rounds, 683 ms for 1500 rounds, and 599 ms for 2000 rounds. The proposed RWMVO with OBL algorithm has obtained higher throughput of 5.68 Kbps and lower energy consumption of 9.05 mW for 500 rounds compared to the existing EASDN (Ahmed 2023) algorithm.

Conclusion

The conclusion exhibits the total performance of the proposed RWMVO with OBL algorithm for edge computing-based multidata collection task using WSN for smart agriculture. The collection of data related to the agriculture field is essential for monitoring the condition of the land. The best sensors nodes were selected from the available resources based on their energy efficiency that retained the energy during the process of multidata collection. The proposed RWMVO with OBL algorithm has obtained higher throughput of 5.68 Kbps and lower energy consumption of 9.05 mW for 500 rounds. The developed RWMVO with OBL algorithm has exhibited the minimal time delay of 795 ms for 500 rounds, 701 ms for 1000 rounds, 683 ms for 1500 rounds, and 599 ms for 2000 rounds compared to the existing EASDN (Ahmed 2023) algorithm. Further, the performance of multidata collection task using WSN for smart agriculture can be improved by incorporating the advanced deep learning (DL)-based optimization algorithm for optimizing the performance of the sensor nodes during the collection of data that save the energy of the sensor nodes.

References

Ahmed S (2023) Energy aware software defined network model for communication of sensors deployed in precision agriculture. Sensors 23(11):5177

Akhter R, Sofi SA (2022) Precision agriculture using IoT data analytics and machine learning. J King Saud Univ Comput Inf Sci 34(8):5602–5618

Aldossary M, Alharbi HA, Anwar Ul Hassan C (2024) Internet of Things (IoT)-enabled machine learning models for efficient monitoring of smart agriculture. IEEE Access 12:75718–75734

Ali MY, Alsaeedi A, Shah SAA, Yafooz WM, Malik AW (2023) Energy efficient data dissemination for large-scale smart farming using reinforcement learning. Electronics 12(5):1248

Bindu LR, Titus P, Dhanya D (2023) Clustered wireless sensor network in precision agriculture via graph theory. Intell Autom Soft Comput 36(2):1435

Chamara N, Bai G, Ge Y (2023) AICropCAM: deploying classification, segmentation, detection, and counting deep–learning models for crop monitoring on the edge. Comput Electron Agric 215:108420

Itoo S, Khan AA, Ahmad M, Idrisi MJ (2023) A secure and privacy-preserving lightweight authentication and key exchange algorithm for smart agriculture monitoring system. IEEE Access 11:56875–56890

Križanović V, Grgić K, Spišić J, Žagar D (2023) An advanced energy–efficient environmental monitoring in precision agriculture using lora-based wireless sensor networks. Sensors 23(14):6332

Makondo N, Kobo HI, Mathonsi TE, Plessis DPD (2024) Implementing an efficient architecture for latency optimisation in smart farming. IEEE Access 12:140502–140526

Raghuvanshi A, Singh UK, Sajja GS, Pallathadka H, Asenso E, Kamal M, Singh A, Phasinam K (2022) Intrusion detection using machine learning for risk mitigation in IoT-enabled smart irrigation in smart farming. J Food Qual 2022(1):3955514

Rosero-Montalvo PD, Gordillo-Gordillo CA, Hernandez W (2023) Smart farming robot for detecting environmental conditions in a greenhouse. IEEE Access 11:57843–57853

Saba T, Rehman A, Haseeb K, Bahaj SA, Lloret J (2023) Trust–based decentralized blockchain system with machine learning using Internet of agriculture things. Comput Electr Eng 108:108674

Sudhakar M, Anne KR (2024) Optimizing data processing for edge–enabled IoT devices using deep learning based heterogeneous data clustering approach. Meas Sens 31:101013

Tseng LM, Chen PF, Wen CY (2023) Design of edge–IoMT network architecture with weight–based scheduling. Sensors 23(20):8553

Wen J, Yang J, Wang T, Li Y, Lv Z (2023) Energy-efficient task allocation for reliable parallel computation of cluster–based wireless sensor network in edge computing. Digit Commun Netw 9(2):473–482

Chapter 13
Recurrent Neural Network with Chaotic Henry Gas Solubility Optimization Algorithm for Predicting Privacy Preservation in Edge Computing

Padmavathi Vurubindi, Sujatha Canavoy Narahari, Naluguru Udaya Kumar, A. Ushasree, and N. Nagalakshmi

Abstract The resource sharing task in the edge computing has become difficult in the recent days due to wide range of access and request in the network. The existing technique tried to reduce the burden in the edge computing to preserve the data privacy against the unauthorized access in the network during the process of resource sharing but failed to reach the expected outcome. The recurrent neural network with chaotic Henry gas solubility optimization (RNN with CHGSO) algorithm is proposed to overcome the existing problem for the unauthorized access during the resource sharing in the edge computing. The recurrent neural network (RNN) predicts the resources availability by interpreting the previous history data of edge computing. Chaotic Henry gas solubility optimization (CHGSO) algorithm finely tuned the hyperparameters of the RNN that advanced the prediction performance of the model for preserving the privacy of edge computing during resource sharing. The developed

P. Vurubindi
Department of Computer Science and Engineering, Chaitanya Bharathi Institute of Technology, Gandipet, Hyderabad, India
e-mail: padmareddyvch@ieee.org

S. C. Narahari
Department of Electronics and Communication Engineering, Sreenidhi Institute of Science and Technology, Ghatkesar, Hyderabad, India
e-mail: sujathac.n@sreenidhi.edu.in

N. U. Kumar (✉)
Department of Electronics and Communication Engineering, Research Centre VTU, Vivekananda Institute of Technology, Belagavi, India
e-mail: joyudaya@gmail.com

A. Ushasree
Department of Computer Science and Engineering, Gokaraju Lailavathi Engineering College, Kukatpally, Hyderabad, India

N. Nagalakshmi
Department of Information Technology, Anurag University, Hyderabad, India
e-mail: nagalakshmiit@anurag.edu.in

G. P. Ramesh et al. (eds.), *Empowering IoT with Distributed Edge Computing*, IoT and Wireless Communication Networks, https://doi.org/10.1007/978-981-96-9455-6_13

RNN with CHGSO algorithm has exhibited better results with accuracy of 0.9624, precision of 0.9143, recall of 0.9087, and f1-score of 0.91114 compared to the existing long short-term memory (LSTM) algorithm.

Keywords Chaotic Henry gas solubility optimization · Edge computing · Privacy preservation · Recurrent neural network · Resource sharing · User request

Introduction

The edge computing refers to the dispersal of computing pattern which summarizes the data storage near to the required location (Shen et al. 2023). The data get processed locally in the edge computing that is nearer to the data source and the users (Simonet-Boulogne et al. 2022). The computational resources such as storage, network bandwidth, and processing power of the edge devices are involved with the resource sharing in the edge computing (Telikani et al. 2023). The distribution of the works across the multiple edges of the devices is known as resource sharing (Quan et al. 2023). There are wide range applications of the edge computing in the fields of smart cities, industries, healthcare and other sectors (Rafique et al. 2023). The resource sharing is mainly depending on the reliable communication between the devices. The securing and privacy preservation of the data is the important task during the process of sharing the resources from one device to other (Yao et al. 2024). The data processed in the edge server should be protected against the irrelevant network access (Qin et al. 2023). The establishment of trust among the users in the edge computing is a difficult task due to the presence of vulnerable activities in the network (Parra-Ullauri et al. 2024). The integrity of the data must be ensured in the edge server to preserve the confidentiality of data. The irrelevant assignment of tasks during resource sharing led to miscommunication in the edge server that reduces the stability of the network (Hegde et al. 2024). The deep learning (DL) model's applications are growing widely in the edge server for the prediction of upcoming availability of resources in the edge computing for the secured resources sharing that preserves the privacy of data.

Zhang et al. (2023) presented a privacy-preserving edge computing using federal learning mechanism. The Paillier homomorphic cryptosystem was used for the verification of user access in the network. The verifiable privacy-preserving federal learning (VPPFL) advanced the performance of the by providing strong key generation for data security in the edge computing. However, the latency in the transmission of data caused interruption in the network where the data packets get lost and reduced the privacy preserving ability of the model. Liu et al. (2022) developed a federal learning-based privacy preservation mechanism for edge computing data. A lightweighted encryption model was incorporated for advanced the parameters of privacy preservation. The convolutional neural network (CNN) was used for the generation of features of the edge computing for reduction of computational consumptions that advanced the performance of preserving the privacy of data transmission in the network. However, the unauthorized access in the network interrupted the verification process that suppressed the privacy preserving ability in the edge computing. Kumar et al. (2022) implemented a blockchain-based privacy preserving mechanism of the edge computing. The attention-based bidirectional long short-term memory (ABiLSTM) classifier was used for the detection of vulnerable activities in the network. The Chameleon swarm optimization (CSO) model optimized the parameters of the ABiLSTM that increased the stability of the privacy preserving mechanism. However, the model failed to secure the data due to lack of improper parameter selection that minimized the security performance in the edge computing.

Jing et al. (2024) presented generative artificial intelligence (AI)-based security for the edge computing data. The blockchain was incorporated for the preserving the data by providing the authentication for the users. The AI model used for collecting the data based on the user's requirements that improved the authentication process of the edge computing that increased the security of the stored data. However, the model struggled to optimize the data due to limited resource that decreased the data security mechanism stability in the edge computing. Rivadeneira et al. (2024) developed an AI-based privacy preservation mechanism in the edge computing. The LSTMclassifier was used for the detecting the available resources in the edge computing for the transmission of data. The human loop-based cyber physical system was used for the collection of data that improved the security of data in the edge server. However, the model failed to identify the data packets due to the network traffic that reduced the stability of edge server security mechanism. Mahmud et al. (2024) presented a federal learning-based intrusion detection mechanism for the preserving the privacy of the edge server. The Federated Averaging (FedAvg) model was developed for the predicting the intrusions present in the network. Based on the iterative weights that increased the security performance of the FedAvg model in the edge server for preserving the data privacy. However, the model performance was affected due to delay in the user verification process that minimized the data security in the edge server. The problems faced by the existing techniques during the process of preserving the privacy in the edge server for resource sharing are given as: The traditional algorithm failed to distribute the resources optimally which utilized huge energy in the edge computing. The unauthorized access in the network led to data packet loss in the network where the information gets vanished. The delay in user verification process also causes the unauthorized access in the network that affects the security of the data stored or transmitted in the edge computing. The contributions of the developed RNN with CHGSO algorithm for privacy preservation in the edge computing during resource sharing is given below as follows:

- The resource focused on preserving the privacy of the edge computing data during the process of resource sharing by reducing the computation time that secures the privacy of data against unauthorized access in the network.
- The recurrent neural network (RNN) algorithm was used of the prediction of resource availability in the edge computing based on the previous history record details that reduces the traffic in the network.

- The chaotic Henry gas solubility optimization (CHGSO) algorithm was incorporated in the model that upgraded the hyperparameters of the RNN algorithm which helps to advance the learning ability of the model for predicting appropriate resource sharing in the edge server that secures and preserve the privacy of data.
- Recurrent neural network with chaotic Henry gas solubility optimization (RNN with CHGSO) algorithm was developed for the optimized prediction of resource sharing performance in the edge server for preserving the data privacy.

The remaining section of the research paper is configured as follows: Section "Proposed Methodology" exhibits the proposed RNN with CHGSO methodology. Section "Experimental Results" illustrates the experimental results. Section "Discussion" displays the discussion, and Section "Conclusion" contains the conclusion, respectively.

Proposed Methodology

The RNN with CHGSO algorithm is proposed for the privacy preservation in the edge computing during the process of resource sharing. The RNN interprets the previous history details of the edge server and then predicts the resource sharing based on the previous history to preserve data privacy. The CHGSO algorithm was incorporated to upgrade the hyperparameters of the RNN that improves the predicting ability of the model that helps for the preservation of data privacy during the resource sharing process. Figure 13.1 exhibits the block diagram of the proposed RNN with CHGSO algorithm for privacy preservation.

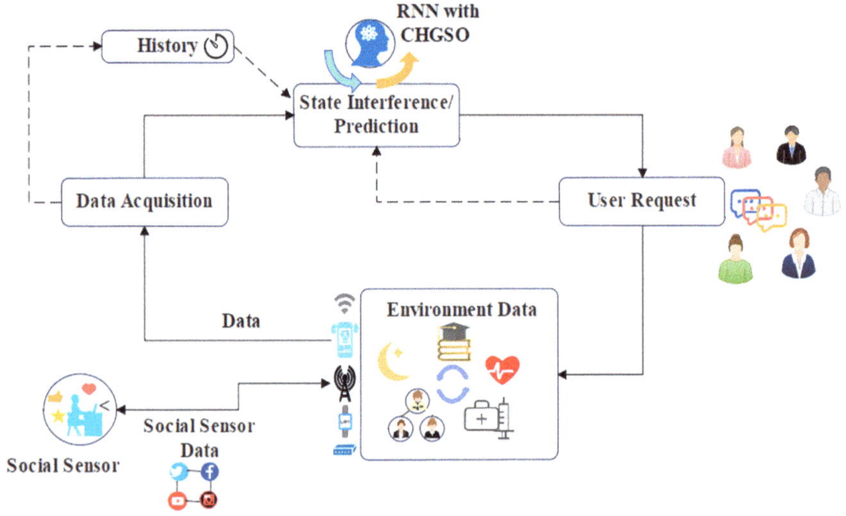

Fig. 13.1 Block diagram of the proposed RNN with CHGSO algorithm for privacy preservation

The Flow of the Privacy Preserving Model in the Edge Computing During Resource Sharing is Given as Follows

Social Sensors and Social Sensor data. These are the user devices gathers the information related to the user activities. The data are processed for minimizing the risk of preserving the privacy during the process of resource sharing.

Environment Data. The data are collected from various platforms like healthcare, industries and other sectors which is processed and then stored the data in the edge server that preserves the data privacy by mitigating centralized storage. The data security is enabled based on the user's privacy mechanisms.

Data Acquisition. The data transferred for the social sensors and environments are all collected in this stage. Initially the data get processed to collects the required or necessary data by filtering the unrelated information that preserves the data privacy before transmission.

State Interference/Prediction. The RNN with CHGSO algorithm is proposed for the prediction of user status in the edge server. The RNN in the edge computing predicts the requirements of the resource sharing that preserves the data and the CHGSO algorithm optimizes the resource sharing performance that helps to preserve the privacy of resource sharing in the edge computing.

History. The RNN model learns the pattern of the user status based on the previous history of edge computing data. The irrelevant exposures of data during the resource sharing in the interface is preserved to secure the data.

User Request. The privacy preservation mechanism ensures that only request data are shared to the user request for the specified details that secures the sensitive information. The user request controls the data availability and sharing in the edge computing.

Output to Users. The gives the details of the predicted users request resource privacy preservation. The aggregated data are shared only when the user request meets the conditions of the privacy preservation.

Proposed LSTM with CHGSO Algorithm

The RNN classifier is used for the prediction of users request in the edge computing for predicting the resource sharing. The neurons in the RNN were interconnected with each other in a directed cycle. The RNN outcomes are depending on proceeding results of the resource sharing. The resources availability in edge computing are predicted using RNN network. The mathematical representation of RNN is given in Eq. (13.1),

$$h_t = f_w(h_t - 1, x_t) \tag{13.1}$$

Here, h_t indicates the novel state for t time, f_w exhibits the function with w vaiable, x_t gives the input vector for t time and $h_t - 1$ shows the older state. The weights are initially allotted using Eq. (13.2),

$$h_t = \tanh(W_{hh}h_1 + W_{xh}x_t) \tag{13.2}$$

The hyperparameters of the RNN algorithm gets finely tuned with the CHGSO algorithm that increases the privacy preserving ability of the model by predicting the appropriate resource sharing in the edge computing.

The HGSO algorithm is used to tune the hyperparameters of RNN by several processing stages. In the initialization stage, the number of gas particles is determined, where the Henry's constant, constant, and the partial gas pressure are allotted using Eqs. (13.3) and (13.4),

$$X_i(t + 1) = X_{\min} + r * (X_{\max} - X_{\min}) \tag{13.3}$$

$$\begin{aligned} H_j(t) &= l_1 * r \\ P_{ij}(t) &= l_2 * r \\ C_j(t) &= l_3 * r \end{aligned} \tag{13.4}$$

Here $X_{\max}$ and $X_{\min}$ give the limitation, r exhibits the random value, X_i shows the location of ith gas particle presents in the gas particle population, $H_j(t)$ exhibits the Henry's constant for jth cluster, $P_{ij}(t)$ provides the partial gas pressure in jth cluster for ith gas particle, l_1, l_2, and l_3 exhibit the constant values. The clustering, evaluation, updating the Henry constant, updating the positions, escape from local optima, and updating the positions of the worst agents are the mechanism of HGSO algorithm. In the clustering phase, the gas particle position is segregated into two divisions of equal gas particles. The Henry and enthalpy constants remain same due to presence of similar gas particle numbers in the clusters. In the evaluation phase, the clusters of the gas particles are assessed on the basis of objective functioning. The assessed gas particles are organized from good results to bad results based on the evaluation. The henry constants are updated using Eq. (13.5),

$$\begin{aligned} H_j(t + 1) &= H_j(t) * \exp\left(-C * \frac{1}{T(t)} - \frac{1}{T^0}\right) \\ T(t) &= \exp\left(-\frac{t}{\text{iter}}\right) \end{aligned} \tag{13.5}$$

Here, $H_j(t)$ exhibits the cluster Henry constant, T provides the temperature, t shows the iteration and iter gives the total iteration value. The gas solubility is upgraded using Eq. (13.6),

$$S_{ij}(t) = K * H_j(t+1) * P_{ij}(t) \tag{13.6}$$

Here, S_{ij} illustrates the solubility for jth cluster for ith iteration, P_{ij} shows the partial gas pressure. The positions of the gas particles are upgraded based on Eq. (13.7),

$$\begin{aligned} X_{ij}(t+1) = X_{ij}(t) + F * r * \gamma * \left(X_{\text{ibest}}(t) - X_{ij}(t)\right) \\ + F * r * a * \left(S_{ij}(t) \times X_{\text{best}}(t) - X_{ij}(t)\right) \end{aligned} \tag{13.7}$$

Here, $\gamma = \beta * \exp\left(-\frac{F_{\text{best}}(t)+\varepsilon}{F_{ij}(t)+\varepsilon}\right)$ and $\varepsilon = 0.05$, X_{ibest} gives the location of the optimal gas particle, γ indicates the ability of the gas particle, a exhibits the gas particle effect, β shows the constant value. The chaotic mechanism is incorporated in the HGSO algorithm for the upgradation of the gas particle positions by finding the fittest optimal explanation for the preservation of privacy in the edge computing during resource sharing using Eq. (13.8),

$$\begin{aligned} X_{ij}(t+1) = X_{ij}(t) + F * r_c * \gamma * \left(X_{\text{ibest}}(t) - X_{ij}(t)\right) \\ + F * r_c * a * \left(S_{ij}(t) \times X_{\text{best}}(t) - X_{ij}(t)\right) \end{aligned} \tag{13.8}$$

Here, r_c gives the chaotic system based generated randomly value. The local optima are escaped in the worst gas particles using Eq. (13.9),

$$N_w(t) = N * (r_c * (c_2 - c_1) + c_1) \tag{13.9}$$

Here $c_1 = 0.1$ and $c_2 = 0.2$, N_w represent the worst gas particle value, the worst position of the gas particles was upgraded using Eq. (13.10),

$$G_{ij}(t+1) = G_{\min} + r_c * (G_{\max} - G_{\min}) \tag{13.10}$$

The CHGSO algorithm tuned the hyperparameters of the RNN that advanced its space searching ability that predicted appropriate resource sharing in the edge computing that preserved the privacy of the data. Further the RNN with CHGSO algorithm undergo evaluation on the basis of existing metrics.

Experimental Results

The proposed RNN with CHGSO method is employed using Python version 3.10 software and the system configuration requirements are operating system (OS) windows 10, processor intel core i7, Random Access Memory (RAM) 16 Giga Byte (GB), Memory 1 Tera Byte (1 TB), and Graphic Processing Unit (GPU) 6 GB. The developed RNN with CHGSO model performance is assessed using existing accuracy, recall, f1-score, and precision metrics. The mathematical representations of the

Table 13.1 Performance of the proposed RNN with CHGSO method

Methodology	Accuracy	Precision	Recall	F1-score
CNN	0.8760	0.8478	0.8517	0.8497
DNN	0.8856	0.8511	0.8675	0.8592
ANN	0.9075	0.8622	0.8721	0.8671
RNN	0.9241	0.8774	0.8867	0.8820
Proposed RNN with CHGSO	0.9624	0.9143	0.9087	0.9114

existing metrics are given from Eqs. (13.11)–(13.14),

$$\text{Precision} = \frac{\text{TP}}{\text{TP} + \text{FP}} \times 100 \tag{13.11}$$

$$\text{Accuracy} = \frac{\text{TP} + \text{TN}}{\text{TP} + \text{TN} + \text{FP} + \text{FN}} \times 100 \tag{13.12}$$

$$\text{Detection Rate(Recall)} = \frac{\text{TP}}{\text{TP} + \text{FN}} \times 100 \tag{13.13}$$

$$F1\text{-Score} = 2 * \left(\frac{\text{Precision} * \text{Recall}}{\text{Precision} + \text{Recall}}\right) \tag{13.14}$$

Here TP exhibits true-positive numeric, TN displays true-negative numeric, FP shows false-positive numeric and FN indicates the false-negative numeric. Table 13.1 illustrates the performance of the proposed RNN with CHGSO method. The developed RNN with CHGSO model has obtained accuracy of 0.9624, precision of 0.9143, recall of 0.9087, and f1-score of 0.9114 compared to the existing algorithms. The CNN, deep neural network (DNN), artificial neural network (ANN), and traditional RNN are the algorithms used for the comparison of the developed RNN with CHGSO model.

Comparative Assessment

The comparative assessment section provides the advancement of the proposed RNN with CHGSO algorithm compared to the existing algorithms. This assessment provides the upgradation of the proposed RNN with CHGSO model for preserving privacy of edge computing during resource sharing compared to existing techniques. The LSTM (Rivadeneira et al. 2024) and FCN (Mahmud et al. 2024) are the two models used for comparison. Table 13.2 illustrates the comparative assessment of the proposed RNN with CHGSO method with the existing algorithms. The developed RNN with CHGSO model has obtained better results with accuracy of 0.9624, precision of 0.9143, recall of 0.9087, and f1-score of 0.9114, respectively.

Table 13.2 Comparative assessment of the proposed RNN with CHGSO model with the existing algorithms

Methodology	Accuracy	Precision	Recall	F1-score
LSTM (Rivadeneira et al. 2024)	0.730	0.671	0.605	0.611
FCN (Mahmud et al. 2024)	0.9153	0.8743	0.8749	0.8746
Proposed RNN with CHGSO	0.9624	0.9143	0.9087	0.9114

Discussion

The discussion provides the information of the drawbacks faced by the existing methods during the process of privacy preservation in the edge computing for resource sharing along with the benefits of the proposed RNN with CHGSO model. The LSTM (Rivadeneira et al. 2024) and FCN (Mahmud et al. 2024) are the existing methods used for the comparing the developed RNN with CHGSO technique. The LSTM (Rivadeneira et al. 2024) failed to predict the traffic in the network during the process of resource sharing that affected the privacy of data and some of the data packets were lost during the sharing process. The FCN (Mahmud et al. 2024) algorithm performance was affected due to delay in the user verification process that led to vulnerable activities in the network and minimized the data security in the edge server. The proposed RNN with CHGSO algorithm overcome the existing problem by optimizing the resource sharing that preserved the privacy of data in the edge computing. The developed RNN with CHGSO model has obtained accuracy of 0.9624, precision of 0.9143, recall of 0.9087, and f1-score of 0.9114 compared to the existing LSTM (Rivadeneira et al. 2024) and FCN (Mahmud et al. 2024) algorithms.

Conclusion

The conclusion exhibits the quick glace of the process used by the developed RNN with CHGSO algorithm in the edge computing during resource sharing with preserving privacy. The RNN algorithm was used for the prediction of resource sharing in the edge computing on the basis of previous history details of the edge computing. The learns and interprets the upcoming resource sharing using history data. The CHGSO algorithm tune the hyperparameters of the model that upgraded the space searching ability of RNN for finding the fittest explanation for the process of resource sharing in the edge computing that preserved the privacy of data against unauthorize access in the network. The developed RNN with CHGSO model has obtained accuracy of 0.9624, precision of 0.9143, recall of 0.9087, and f1-score of 0.9114 compared to the existing LSTM (Rivadeneira et al. 2024) and FCN (Mahmud et al. 2024) algorithms. Further, the edge computing privacy preserving performance for the resource sharing can be advanced by incorporating improved optimization-based deep learning techniques for predicting the traffic in the edge computing during resource sharing process.

References

Hegde C, Kiarashi Y, Rodriguez AD, Levey AI, Doiron M, Kwon H, Clifford GD (2024) Indoor group identification and localization using privacy-preserving edge computing distributed camera network. IEEE J Indoor Seamless Position Navig 2:51

Jing Z, Xu X, Gu C, Zhang Y, Shu Q (2024) A security-enhanced advertising platform based on blockchain and edge computing in generative AI. Appl Artif Intell 38(1):2340395

Kumar K, Mahilraj J, Swathi D, Rajavarman R, Zeebaree SR, Zebari RR, Rashid ZN, Alkhayyat A (2022) Privacy preserving blockchain with optimal deep learning model for smart cities. Comput Mater Contin 73(3):5299–5314

Liu Z, Gao Z, Wang J, Liu Q, Wei J (2022) PPEFL: an edge federated learning architecture with privacy-preserving mechanism. Wirel Commun Mob Comput 2022(1):1657558

Mahmud SA, Islam N, Islam Z, Rahman Z, Mehedi ST (2024) Privacy-preserving federated learning-based intrusion detection technique for cyber-physical systems. Mathematics 12(20):3194

Parra-Ullauri JM, Madhukumar H, Nicolaescu AC, Zhang X, Bravalheri A, Hussain R, Vasilakos X, Nejabati R, Simeonidou D (2024) KubeFlower: A privacy-preserving framework for Kubernetes-based federated learning in cloud–edge environments. Futur Gener Comput Syst 157:558–572

Qin J, Zhang X, Liu B, Qian J (2023) A split-federated learning and edge-cloud based efficient and privacy-preserving large-scale item recommendation model. J Cloud Comput 12(1):57

Quan G, Yao Z, Chen L, Fang Y, Zhu W, Si X, Li M (2023) A trusted medical data sharing framework for edge computing leveraging blockchain and outsourced computation. Heliyon 9(12):e22542

Rafique W, Khan M, Khan S, Ally JS (2023) Securemed: a blockchain-based privacy-preserving framework for internet of medical things. Wirel Commun Mob Comput 2023(1):2558469

Rivadeneira JE, Borges GA, Rodrigues A, Boavida F, Silva JS (2024) A unified privacy preserving model with AI at the edge for human-in-the-loop cyber-physical systems. Internet of Things 25:101034

Shen Y, Shen S, Li Q, Zhou H, Wu Z, Qu Y (2023) Evolutionary privacy-preserving learning strategies for edge-based IoT data sharing schemes. Digit Commun Netw 9(4):906–919

Simonet-Boulogne A, Solberg A, Sinaeepourfard A, Roman D, Perales F, Ledakis G, Plakas I, Sengupta S (2022) Toward blockchain-based fog and edge computing for privacy-preserving smart cities. Front Sustain Cities 4:846987

Telikani A, Shahbahrami A, Shen J, Gaydadjiev G, Lin JCW (2023) An edge-aided parallel evolutionary privacy-preserving algorithm for Internet of Things. Internet of Things 23:100831

Yao A, Pal S, Li X, Zhang Z, Dong C, Jiang F, Liu X (2024) A privacy-preserving location data collection framework for intelligent systems in edge computing. Ad Hoc Netw 161:103532

Zhang J, Liu Y, Wu D, Lou S, Chen B, Yu S (2023) VPFL: a verifiable privacy-preserving federated learning scheme for edge computing systems. Digit Commun Netw 9(4):981–989

Chapter 14
Graph Neural Network-Based Long Short-Term Memory with Common Vulnerability Scoring System for Security Threat Detection in Edge Computing

H. Manoj T. Gadiyar, K. Arjun, Mohan Ramachandra Naik, M. Bharathraj Kumar, and Gurusiddayya Hiremath

Abstract The rapid growth in the technology advancement has increased the demand for smart device. The edge computing applications are rising in the due to its vast benefits of storage and accessibility. As the users are increasing, the vulnerable activities are also rising in the network that cause security threats for the edge computing data. The existing methods tried to detect the security threats in the network to secure the data but did not get the expected outcome due to poor representation of node networks. The graph neural network-based long short-term memory with common vulnerability scoring system (GNN-LSTM with CVSS) model was developed for the detection of security threats in the edge computing. The graph neural network (GNN) represented the node network graph to analyze the threats. The ResNet-50 extracted the important details from the node network graph. The long short-term memory (LSTM) interpretated the patterns of the extracted features for the security threat detection in the edge computing. The developed GNN-LSTM with CVSS has increased the security threat detection performance. The proposed GNN-LSTM with CVSS model has obtained greater results of 94.27% accuracy and 93.73% f1-score compared to the existing distilled-bidirectional encoder representations from transformer (DistilBERT) model.

H. M. T. Gadiyar
Department of Information Science and Engineering, Canara Engineering College, Mangaluru, India

K. Arjun
Department of Artificial Intelligence and Machine Learning, Canara Engineering College, Mangaluru, India

M. Ramachandra Naik (✉) · M. Bharathraj Kumar
Department of Electronics and Communication Engineering, Shri Dharmasthala Manjunatheshwara Institute of Technology, Ujire, India
e-mail: mohannaik@sdmit.in

G. Hiremath
Department of CSE (Artificial Intelligence and Machine Learning), Sahyadri College of Engineering & Management, Mangaluru, India

G. P. Ramesh et al. (eds.), *Empowering IoT with Distributed Edge Computing*, IoT and Wireless Communication Networks, https://doi.org/10.1007/978-981-96-9455-6_14

Keywords Common vulnerability scoring system · Edge computing · Graph neural network · Long short-term memory · ResNet-50 · Security threat

Introduction

The devices that are situated near to the edges of the devices or user ends is generally known as edge computing (Shin et al. 2022). The edge devices, gateways, servers, network infrastructure, and backend infrastructure are the major components of the edge computing that manages the data (Pathak et al. 2024). The edge devices are the devices that are present at the edges of the network to gather the data from the environment, the data get processed here before transmitting to the nearer gateways (Jiao et al. 2024). The data gets processed and stored closer to the devices that produce the data that helps in minimizing the network bandwidth and response time (Nebbione and Calzarossa 2023). The traveling of the data in the network determines the computational performance of the edge computing (Shafee et al. 2023). The common vulnerability scoring system (CVSS) is a framework that provides the numeric score for the analysis of software vulnerabilities (Sharaf et al. 2024). The security threats in the networks are prioritized based on the CVSS by analyzing their impacts on the network (Shitharth et al. 2023). Each device present in the edge computing generates the vulnerability like unpatched software, outdated firmware and other malwares in the network (Epiphaniou et al. 2023). The CVSS focus on the critical vulnerabilities present in the edge computing devices to secure the data (Zhang et al. 2023). The impacts of the vulnerabilities for the security of network are assessed based on the score of the CVSS. The contribution of the proposed GNN-LSTM with CVSS model for the detection of security threats in the edge computing is given below as follows:

- The research aims in the advancement of security threat detection in the edge computing by integrating the deep learning algorithms with the common vulnerability scoring system (CVSS) model.
- The graph neural network (GNN) algorithm was used for the advancement of node network graph representation to get appropriate visualization of the networks that helps for the analysis of security threats.
- The ResNet-50 algorithm was used for the extraction of significant details from the network graphs that simplifies the features of the network by minimizing the dimensionalities.
- The long short-term memory (LSTM) classifier was used for the appropriate detection of security threats in the edge computing based on the previous history data of the networks.
- The graph neural network-long short-term memory (GNN-LSTM) with common vulnerability scoring system (CVSS) upgraded the detection of security threats in the edge computing by providing the priorities to the vulnerable activities in the network.

Literature Survey

The literature survey provides the details of the benefits and drawbacks of the existing methodologies used for the security threat detection in the edge computing. Adamos et al. (2024) presented an enhanced cyberattacks physical system for the risk assessment. The isomorphic graph of the CPS process model was created and used graph algorithms and then the network was analyzed by testing the cyberattacks in the network. The model resisted to vulnerable activities in the network that advance the security of the edge computing. However, the model failed to identify the performance of the task in the edge computing that increased security threats in the edge server. Kim et al. (2023) developed a moving target defense strategy based on time to detect the security threats in the edge computing. The Bayesian attack graph (BAG) was used for the analysis of network movement analysis. The common vulnerability scoring system (CVSS) was used with the BAG for the assessment of security risk in the network that improved the security of the edge computing against the vulnerable threats. However, the model struggled to secure the data due to unauthorized access in the network that suppressed performance of security mechanism in the network. Sun et al. (2024) implemented a risk assessment mechanism based on graph embedding technique for the prevention of security threats in the network. The node to vector technique was applied for the analysis of vulnerabilities in the network. The weights were synthesized by using cyber-physical graph model that advanced the risk assessment of the security threats in the network. However, the model failed to handle the multiple tasks assigned in the network due to latency in the data transmission that suppressed the security of the edge network.

Kazeminajafabadi and Imani (2023) presented a BAG-based network threat detection mechanism. The adaptive resource monitoring policy was used for monitoring the activities of the nodes in the network. The BAG mechanism identified the hidden Markov model that increased the nodes monitoring mechanism performance in the network that increased the security of the network against multiple vulnerable threats. However, the stability of the security mechanism was affected due to loss of data packets during the transmission process that increased the threats activities in the network. Costa et al. (2022) developed natural learning processing (NLP)-based CVSS prediction mechanism for securing the edge computing. The distilled-bidirectional encoder representations from transformer (DistilBERT) technique was used for predicting the threats in the network. The security threats in the network were analyzed based on the Shapley value that advanced the prediction performance of the model to get the appropriated security threat. However, the model learnt the noise generated during the training process that suppressed the security threat predicting ability of the model. Liu et al. (2024) implemented a security threat assessment model based on blockchain mechanism in the edge computing. The property graph model was used for the visualization of network activities. The Ant Colony Optimization of Key Node (KNACO) was used for the optimized the performance of the network to identify the security threats based on the weights of the nodes that improved assessment of security threats in the network. However, the model did not handle the model consumed huge energy resources for the assessment the network threats that minimized the stability of the edge computing. The existing techniques faced various types of drawbacks during the detection of security threats in the edge computing which are given here: The stability of the security mechanism was affected due to loss of data packets during the transmission process that increased the threats activities in the network. The traditional model failed to identify the performance of the task in the edge computing that increased security threats in the edge server. The model struggled to secure the data due to unauthorized access in the network that suppressed performance of security mechanism in the network. The model learnt the noise generated during the training process that suppressed the security threat predicting ability of the model in the edge computing. The model failed to handle the multiple tasks assigned in the network due to latency in the data transmission that suppressed the security of the edge network.

Proposed Methodology

The GNN-LSTM with CVSS model is proposed for the detection of cyber threats in the edge computing. The procedure of the detection model is executed in four stages like edge computing for data collection, graphical representation of edge nodes using GNN algorithm, feature extraction using ResNet-50 algorithm and then detection of the security threats using LSTM algorithm, respectively. After the data collection in the edge computing, the node networks are visualized in the graph form using GNN algorithm, further the ResNet-50 was used for the extraction of significant details from the network graph that increases the security threat detection ability of the LSTM model. Figure 14.1 indicates the block diagram of the proposed GNN-LSTM with CVSS methodology.

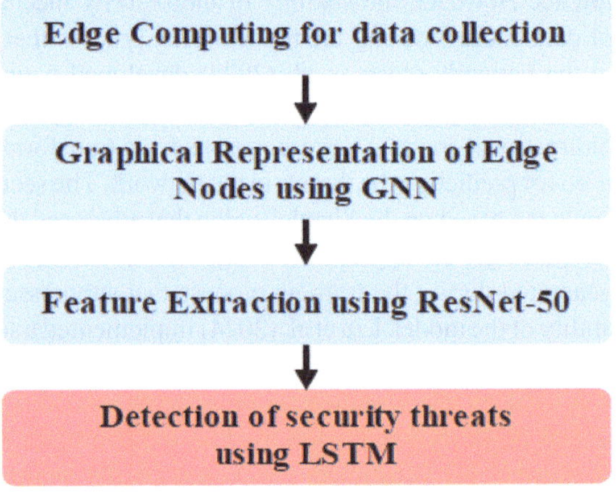

Fig. 14.1 Block diagram of the proposed GNN-LSTM with CVSS methodology

Edge Computing for Data Collection

The data of various devices such as IoT sensors, edge nodes, and smart devices are gathered in the edge computing. The data are collected locally to reduce the latency and bandwidth of the data transmission that boost the processing of data present at the edges. This helps to secure the data by monitor the activities of the edge computing. The edge nodes are further graphically represented to get the visualization of the network for analyzing the security threats.

Graphical Representation of Edge Nodes Using GNN

After the gathering of data, the edge nodes of the devices are represented using GNN model to analyze the node activities. The GNN (Kosasih et al. 2024) is used for analyzing the dependencies of the edge nodes and also interprets the pattern of their communication. The GNN is applied for each node in the network to get aggregated information. The mathematical representation of the data aggregation is given in Eq. (14.1),

$$h_u^{k+1} = \text{UPDATE}^k\left(h_u^k \sum_{r \in R} \text{AGGREGATE}_r^k\left(h_v^k, v \in N_r(u)\right)\right) \tag{14.1}$$

The weight-based activation function is adapted in the GNN network to increase the representation of the nodes. The mathematical representation is given based on Eq. (14.2),

$$h_u^{k+1} = h_u^k \sum_{v \in N_t(u)} W_r^k h_v^k \tag{14.2}$$

Here h_u^k and h_v^k exhibits the embedding of nodes in the GNN layer, $N_r(u)$ gives the set of neighbor node that are connected. The nodes are connected to the edges using Eq. (14.3),

$$f(u, r, v) = \sigma\left(\left(h_u^k\right)^T R_r h_v^k\right) \tag{14.3}$$

Here, R_r provides the specified matrix of edges, σ gives the sigmoid function, and $f(u, r, v)$ shows the dot products of the nodes. The GNN captures the spatial temporal dependencies of the network that increase the decision-making ability of the model. After the completion of node representation, the ResNet-50 model is used for the extraction of significant features from the graphs.

Feature Extraction Using ResNet-50

The graphs of the nodes are taken for the extraction of significant features to advance the representation of features. The ResNet-50 (Wang et al. 2024) algorithm is used of the extraction of important details from the graphs. The residual block present in the ResNet-50 algorithm contain convolutional layers. Each residual block adapted a skip connector that transforms the data to the proceeding layer. The ResNet-50 algorithm extract the important details such as textural and high semantic details. The low-level features and higher-level information's are captured by the ResNet-50 algorithm. The gradient vanishing issue is reduced due to incorporation of residual connector that simplifies the model training. The features are represented using Eq. (14.4),

$$X_{(l+1)} = F(X_l) + W_s \cdot X_1 \tag{14.4}$$

Here, X_l gives the input feature map, $X_{(l+1)}$ exhibits the output feature map, $F(X_l)$ gives the dimension of the projected matrix, and W_s indicates the projected matric of the matching dimension. The extracted features are further transferred to the LSTM model for the detection of security threats in the network.

Detection of Security Threats Using LSTM

The extracted features are the input for the detection stage where the LSTM (Hnamte et al. 2023) model is used for the security threat detection. The LSTM has three gates like input, output and forget gate for the interpretation of the feature details. The long-range dependencies of the node network features are recognized by the LSTM that increased the security threat detection ability of the model in the edge computing. The mathematical representation of the three gates is given below for Eqs. (14.5)–(14.10)

$$e_t = \sigma_h(W_e x_t + P_e I_{t-1} + a_e) \tag{14.5}$$

$$j_t = \sigma_h\left(W_j x_t + P_j I_{t-1} + a_j\right) \tag{14.6}$$

$$p_t = \sigma_h\left(W_p x_t + P_p I_{t-1} + a_p\right) \tag{14.7}$$

$$d_t = \sigma_h(W_d x_t + P_d I_{t-1} + a_d) \tag{14.8}$$

$$d_t = e_t o e_{t-1} + j_t o \overline{d}_t \tag{14.9}$$

$$L_t = p_t o \sigma_t(d_t) \tag{14.10}$$

Here, e_t indicates the forget gate activation vector, L_t exhibits the hidden state activation vector, x_t shows the dimensional space, j_t provides the input gate activation vector, d_t gives the input cell activation vector, and W matrix weight respectively. The LSTM learns the patterns of the extracted features and then detects the security threats in the edge computing based on the previous history records of the edge network. The LSTM get integrated with the CVSS for upgrading the recognition of the vulnerabilities present in the edge computing.

Integration of LSTM with CVSS. The software threats are detected based on the scores assigned by the CVSS that indicates the vulnerability characteristics. The sensitiveness of the threats is categorized using CVSS score and provides alert or security to the network. The CVSS helps for reducing the latency in the edge computing providing the security against the threats. The vulnerabilities in the edge computing are assessed in the structured manner that provides the automated security to the threats in the network. The Integration of LSTM with CVSS advanced the security threat detection ability in the edge computing that helps in the mitigation of security threats and protects the data. After the completion of security threat detection in the edge computing, the GNN-LSTM with CVSS model is assessed to get analyze its performance toward the security threat detection in the edge computing.

Experiment Results

The proposed GNN-LSTM with CVSS model is implemented in the MATLAB software of version R2020b and the configuration of system comprises the processor intel core i7, Operating System (OS) Windows 10, Random Access Memory (RAM) 16 Giga Byte (GB), Memory 1 Tera Byte (TB), Graphic Processing Unit (GPU) 6 GB, respectively. The developed GNN-LSTM with CVSS model is assessed using accuracy, Balanced Accuracy (BA) and f1-score metrics. The mathematical representation of the performance metrics is given in Eqs. (14.11) and (14.12),

$$\text{Accuracy} = \frac{\text{TP} + \text{TN}}{\text{TP} + \text{TN} + \text{FP} + \text{FN}} \times 100 \tag{14.11}$$

$$F1\text{-Score} = \frac{2\text{TP}}{2\text{TP} + \text{FP} + \text{FN}} \times 100 \tag{14.12}$$

Here TP exhibits the true positive numeric, TN provides true negative numeric, FP indicates false-positive numeric, and FN shows the false-negative numeric. Table 14.1 illustrates the performance of the proposed GNN-LSTM with CVSS model for the detection of security threats in the edge computing. The gated recurrent unit (GRU), artificial neural network (ANN), convolutional neural network (CNN), and traditional LSTM are used for the comparison of proposed GNN-LSTM with CVSS model and has obtained better results with accuracy of 94.27% and fi-score of 93.73%, respectively.

Table 14.1 Performance of the proposed GNN-LSTM with CVSS model

Methodology	Accuracy (%)	F1-score (%)
GRU	91.76	90.56
ANN	92.03	91.25
CNN	92.66	92.19
LSTM	93.54	92.86
GNN-LSTM with CVSS	94.27	93.73

Table 14.2 Comparative assessment of the proposed GNN-LSTM with CVSS and DistilBERT (Costa et al. 2022) model

Methodology	Accuracy (%)	F1-score (%)
DistilBERT (Costa et al. 2022)	93.01	92.98
GNN-LSTM with CVSS	94.27	93.73

Comparative Assessment

The comparative assessment is essential to know the upgradation of the proposed GNN-LSTM with CVSS model compared to the existing methods used for the detection of security threats in the edge computing. Table 14.2 exhibits the comparative assessment of the proposed GNN-LSTM with CVSS model and DistilBERT (Costa et al. 2022) model. The developed GNN-LSTM with CVSS model has displayed better results of 94.27% accuracy and 93.73% f1-score, respectively.

Discussion

The discussion provides the details of the existing methods drawbacks for the detection of security threats in the edge computing along with the advantages of the proposed GNN-LSTM with CVSS methodology. The DistilBERT (Costa et al. 2022) model failed to identify the appropriate threats in the network due to lack of proper graph node representation which affected the learning ability of the model and reduced the accuracy of the security threat detection model. The analysis of node networks failed due to poor representation of the network features that suppressed the performance of detecting the security threats in the edge computing. The traditional models struggled with the interpreting the patterns of the security threats due to limited feature searching ability that affected the performance of security threat detection. The model failed to handle the multiple tasks assigned in the network due to latency in the data transmission that suppressed the security of the edge network. The developed GNN-LSTM with CVSS has overcome the existing problem by improving the visualization of node network graph and detected the threats in the network based on the previous history data. The GNN increased the visualization effect of the node networks, then the ResNet-50 algorithm extracted the significant details from the graph and then the LSTM detected the security threats in the edge computing. The proposed GNN-LSTM with CVSS model has obtained greater results of 94.27% accuracy and 93.73% f1-score compared to the existing DistilBERT (Costa et al. 2022) model.

Conclusion

The conclusion shows the overview of the process involved in the edge computing for the detection of cyber threats in the network. The GNN-LSTM with CVSS model is proposed for the cyber threat detection in the edge computing. The data was collected and stored in the edge computing is used for the analysis of security threats. The GNN algorithm graphically represents the node network that helps increased the visualization effects of the network. The ResNet-50 algorithm extracted significant details from the node network graph that simplifies the complexity of the features. The LSTM algorithm interpreted the patterns of the extracted features for the detection of security threats in the edge computing. The GNN-LSTM model detected the security threats based on the previous history data details. The developed GNN-LSTM with CVSS model has shown improved results of accuracy 94.27% and f1-score 93.73% compared to the existing DistilBERT (Costa et al. 2022) model. Further the performance of the edge computing security threat detection performance can be upgraded by the incorporation of modified deep learning techniques with advanced feature representation mechanism to increase the security threat detection performance.

References

Adamos K, Stergiopoulos G, Karamousadakis M, Gritzalis D (2024) Enhancing attack resilience of cyber-physical systems through state dependency graph models. Int J Inf Secur 23(1):187–198

Costa JC, Roxo T, Sequeiros JB, Proenca H, Inacio PR (2022) Predicting CVSS metric via description interpretation. IEEE Access 10:59125–59134

Epiphaniou G, Hammoudeh M, Yuan H, Maple C, Ani U (2023) Digital twins in cyber effects modelling of IoT/CPS points of low resilience. Simul Model Pract Theory 125:102744

Hnamte V, Nhung-Nguyen H, Hussain J, Hwa-Kim Y (2023) A novel two-stage deep learning model for network intrusion detection: LSTM-AE. IEEE Access 11:37131–37148

Jiao J, Li W, Guo D (2024) The vulnerability relationship prediction research for network risk assessment. Electronics 13(17):3350

Kazeminajafabadi A, Imani M (2023) Optimal monitoring and attack detection of networks modeled by Bayesian attack graphs. Cybersecurity 6(1):22

Kim H, Hwang E, Kim D, Cho JH, Moore TJ, Nelson FF, Lim H (2023) Time-based moving target defense using Bayesian attack graph analysis. IEEE Access 11:40511–40524

Kosasih EE, Margaroli F, Gelli S, Aziz A, Wildgoose N, Brintrup A (2024) Towards knowledge graph reasoning for supply chain risk management using graph neural networks. Int J Prod Res 62(15):5596–5612

Liu Y, Pan L, Chen S (2024) A hierarchical blockchain-enabled security-threat assessment architecture for IoV. Digit Commun Netw 10(4):1035–1047

Nebbione G, Calzarossa MC (2023) A methodological framework for AI-assisted security assessments of active directory environments. IEEE Access 11:15119–15130

Pathak V, Singh K, Khan T, Shariq M, Chaudhry SA, Das AK (2024) A secure and lightweight trust evaluation model for enhancing decision-making in resource-constrained industrial WSNs. Sci Rep 14(1):28162

Shafee AA, Mahmoud MM, Srivastava G, Fouda MM, Alsabaan M, Ibrahem MI (2023) Detection of distributed denial of charge (DDOC) attacks using deep neural networks with vector embedding. IEEE Access 11:75381–75397

Sharaf SA, Ragab M, Albogami N, Al-Malaise Al-Ghamdi A, Sabir MF, Maghrabi LA, Ashary EB, Alaidaros H (2024) Advanced mathematical modeling of mitigating security threats in smart grids through deep ensemble model. Sci Rep 14(1):23069

Shin GY, Hong SS, Lee JS, Han IS, Kim HK, Oh HR (2022) Network security node-edge scoring system using attack graph based on vulnerability correlation. Appl Sci 12(14):6852

Shitharth S, Alshareef AM, Khadidos AO, Alyoubi KH, Khadidos AO, Uddin M (2023) A conjugate self-organizing migration (CSOM) and reconciliate multi-agent Markov learning (RMML) based cyborg intelligence mechanism for smart city security. Sci Rep 13(1):15681

Sun S, Huang H, Payne E, Hossain-McKenzie S, Jacobs N, Poor HV, Layton A, Davis K (2024) A graph embedding-based approach for automatic cyber-physical power system risk assessment to prevent and mitigate threats at scale. IET Cyber Phys Syst Theory Appl 9(4):435–453

Wang L, Ji W, Wang G, Feng Y, Du M (2024) Intelligent design and optimization of exercise equipment based on fusion algorithm of yolov5-resnet 50. Alex Eng J 104:710–722

Zhang S, Su X, Han Y, Du T, Shi P (2023) Application research on two-layer threat prediction model based on event graph. Comput Mater Contin 77(3):3994–4023

Chapter 15
Embedded Bidirectional Encoder Representations from Transformers with Regularized Random Forest for Detection of Authentication and Authorization in the Edge Devices

N. V. Babu, Chikkalwar Sudha Rani, R. Rana Veer Samara Sihman Bharattej, and P. Kiran Kumar Reddy

Abstract The process of ensuring the performance and action of the edge devices to access the resources is secured by authenticating and authorizing the devices. As the devices are increasing due to its huge demand, the unauthorized access in the network is increasing that affects the security of the edge device data. The existing models tried to detect the authorized and unauthorized access in the network for reducing the vulnerable activities but failed to handle the dimensionalities of the features that reduced the security of the system. The embedded bidirectional encoder representations from transformers with regularized random forest (EBERT-RRF) model is developed to overcome the existing problem for the detection of authenticating and authorizing the edge devices. The embedded bidirectional encoder representations from transformers (EBERT) model extracted the significant information from the edge device data that increased the training ability of the model by minimizing the overfitting problem. The regularized random forest (RRF) classifier was used for the detection of authentication and authorization of edge devices. The developed EBERT-RRF model has obtained better results of f1-score of 0.859, accuracy of 0.925, recall

N. V. Babu
Department of Electronics and Communication Engineering, SJB Institute of Technology, Bangalore, India
e-mail: nvbabu@sjbit.edu.in

C. S. Rani
Department of Computer Science and Informatics, University College of Engineering and Technology, Mahatma Gandhi University, Nalgonda, India

R. Rana Veer Samara Sihman Bharattej
Doctorate of Business Administration, National Louis University, Tampa, FL, USA

P. Kiran Kumar Reddy (✉)
Department of Computer Science and Engineering (AIML), MLR Institute of Technology, Hyderabad, India
e-mail: kiran.penubaka@gmail.com

G. P. Ramesh et al. (eds.), *Empowering IoT with Distributed Edge Computing*, IoT and Wireless Communication Networks, https://doi.org/10.1007/978-981-96-9455-6_15

of 0.853, and precision of 0.867 compared to the existing extreme gradient boosting (XGB).

Keywords Authentication and authorization · Edge devices · Embedded bidirectional encoder representations from transformers · Internet of Things · Min–max normalization · Regularized random forest

Introduction

The hardware that performs entry or exit point in a network that links the physical world and the database is known as edge devices (Sharadqh et al. 2023). The data collection, data processing, and security are the function of the edge devices and stored in the cloud computing (Mishra et al. 2024). The sensor is used for data gathering and processing. The real time data are collected that increased the customer experiences and product quality (Ajao and Apeh 2023). The collected data are further transmitted to the edge computing where the data get processed. The edge devices present the security to the data against the unauthorized access in the network by controlling the incoming and outgoing traffic in the network (Tan 2023). The embedding of sensors, software, and other technologies for the exchange of data from one system to another system using internet is generally known as Internet of Things (IoT) (Zaidi et al. 2022). The devices in the IoT are the physical objects that contains the sensors, connectivity modules (Almaiah et al. 2022). The sensor is responsible of connecting the environment data and then the data get processed in the edge devices (Ahmadi et al. 2022). The authorization and authentication of the edge devices is essential to secure the collected data. The machine learning (ML) models are used for the recognition of continuous behavioral patterns of the network for the user identity verification (Razzaq et al. 2023). The models learn from the previous history data of the network to identify the authorization of the edge devices (Khashan and Khafajah 2023). The contribution of the proposed EBERT-RRF algorithm for the detection of authenticating and authorizing of the edge devices are given below as follows:

- The research focused presenting the authentication and authorization detection model for reducing the unauthorized access in the edge devices to secure the privacy of stored data in the network.
- The min–max normalization technique was used in the preprocessing stage for rescaling the various ranges of edge device data into standard range that increases the regularization of the model.
- The embedded bidirectional encoder representations from transformers (BERT) algorithm was used of the extraction of significant details of edge devices by converting the data into vector form where the dimensionality of the features gets reduced and advanced the model training ability.

- The regularized random forest (RRF) classifier was used for the appropriate detection of the authentication and authorization of edge devices that helps to increase the security mechanism of the network.
- The embedded bidirectional encoder representations from transformers with regularized random forest (EBERT-RRF) algorithm is developed to detect authentication and authorization of edge devices appropriately to mitigate the irrelevant access in the network that increases the data security.

The rest of the sections present in this research paper is structured as follows: Section "Literature Survey" exhibits the literature survey. Section "Proposed Methodology" displays the proposed EBERT-RRF methodology. Section "Experimental Results" shows the experimental results. Section "Discussion" consists of discussion, and Section "Conclusion" indicates the conclusion.

Literature Survey

This section provides the information of the existing methods used for the detection of authenticating and authorizing of edge devices in the network along with their benefits and drawbacks. Saba et al. (2023) presented a trust based blockchain mechanism for secured authentication of edge computing for IoT data. The blockchain was incorporated for the assurance of data verification and authentication for protecting the privacy of the data. The multivariable linear regression (LR) model was used for the analysis of IoT data and a strong key was generated for authentication of the user access in the network that increased the security of the edge computing. However, the performance of the model was affected due to the traffic in the network that decreased the security of the edge computing. Loconte et al. (2024) developed a federated learning (FL)-based edge computing for IoT data security. The authentication mechanism and the encrypted communication for securing the data in the edge computing. The federated learning was used for the extension of the nodes that increased the training of the model. The encrypted communication was used only for the authorized access that upgraded the edge computing security. However, the performance of the authentication system was affected due limited resisting capability of the model for multiple attacks in the network that minimized the security of the IoT data. Mazzocca et al. (2024) implemented an Internet of Thing Angle (IOTA)-based federal learning mechanism for authentication and authorization of the user's access in the network. The FL at the Edge through the IOTA TAngle (FETA) protocol was used for the authorization of network access and normalize the IoT data. The incorporation of blockchain for data storage and authentication of the users that advanced the safety of edge device data. However, the protection of the edge device data was suppressed due to poor feature interpreting ability of the model that reduced the performance of authentication mechanism in the network.

Ehsan et al. (2024) presented an extreme gradient boosting (XGB)-based attack detection mechanism for the user resource access in the edge devices. The XGB was integrated with the smart contract for the securing the IoT data. The term frequency-inverse document frequency (TF-IDF) was used for the mining of significant details of the user access in the network and the XGB model predicted the unauthorized access in the network that increased the security of IoT data. However, the model interpreted the noises generated during the raining process that led to misclassification of user access details and minimized the authentication performance in the network. Kumar et al. (2023) developed a blockchain-based secure data transmitting mechanism with authorized access in the network. The Deep Sparse AutoEncoder (DSAE) method was used for the extraction of important details of the user's access in the network to analyze the authentications. The bidirectional long short-term memory (BiLSTM) algorithm interpreted the patterns of the extracted features for the identification of unauthorized network access in the edge devices that increased the stability of security model. However, the model failed to handle the dimensionalities of the features that suppressed the performance of the authorization verification mechanism in the network. Sadique et al. (2023) implemented a distributed identity managing mechanism in the IoT devices for the verification of authorized network access. The distributed ledger technology-local identity provider (DLT-LIdP) was used for the identity verification that increased the authenticating and authorizing action of the user access in the edge devices. However, the performance of the model was affected by the multiple access in the network that delayed the user verification process in the network. The existing methods faced various problems during the detection of authentication and authorization of the edge devices which are given below as follows: The traditional model performance was affected due to the traffic in the network that decreased the security of the edge computing. The security of the edge device data was suppressed due to poor feature interpreting ability of the model that reduced the performance of authentication mechanism in the network. The existing model interpreted the noises generated during the raining process that led to misclassification of user access details and minimized the authentication performance in the network. The model failed to handle the dimensionalities of the features that suppressed the performance of the authorization verification mechanism in the network. The performance of the model was affected by the multiple access in the network that delayed the user verification process in the network.

Proposed Methodology

The EBERT-RRF is the model proposed for the detection of authenticating and authorizing in the edge devices. The smart contrast data like solidity code, bytecode, and opcode are used as input data for the model. The min–max normalization is used in the preprocessing stage to rescale the different ranges of data to standard form. The preprocessed data get further transferred to the feature extraction phase where the EBERT algorithm extracted the significant details from the edge device data. The extracted feature is further fed to the RRF classifier for the detection of authenticating and authorizing in the edge devices in the network. Figure 15.1 indicates the block

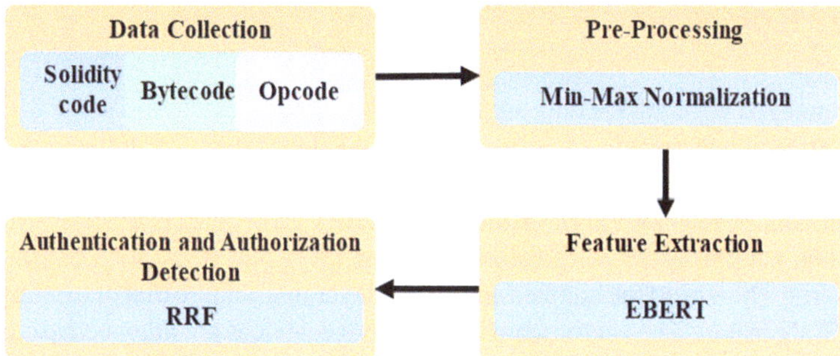

Fig. 15.1 Block representation of the proposed EBERT-RRF for the detection authenticating and authorizing in the edge devices

representation of the proposed EBERT-RRF for the detection of authenticating and authorizing in the edge devices.

Data Collection

The data are collected from various edge devices which are openly available for the access. The source code, bytecode, and opcode are smart contrast codes used for the operation in the IoT that gives the information of authenticating and authorizing in edge devices are the data (Ehsan et al. 2024). These IoT data are the input for the preprocessing stage where the min–max normalization techniques are used for the scaling of data ranges.

Preprocessing

The collected data of the edge devices are the input for the preprocessing stage where the data is rescaled for the fixed range value. The min–max normalization (Alruwaili et al. 2023) technique is used in the preprocessing stage for normalizing the data that increases the training of the model. The mathematical representation of the min–max normalization is shown in Eq. (15.1),

$$X_{\text{norm}} = \left(\frac{X - X_{\min}}{X_{\max} - X_{\min}}\right) \tag{15.1}$$

Here X_{norm} provides the normalize data, X gives the current data, $X_{\max}$ and $X_{\min}$ exhibits the maximum and minimum data value, respectively. After completion of data normalization, the data is further transmitted to the feature extraction stage where the significant details of the edge devices are extracted using EBERT algorithm.

Feature Extraction

The normalized edge device data are the input for feature extraction stage where the important details of the edge devices are taken for the analysis of authentication and authorization. The EBERT (Yu et al. 2024) method is used for the extraction of edge device details. A new layer is embedded in the BERT algorithm. The relevant features of the data get embedded with the help of new layer that transforms the data into vector form. The size of the adapted embedded layer is similar to that of existing layer of BERT algorithm. The vector features of the edge devices are mapped based on the similar feature details that provides high-dimensional representation of the feature vector, indicating the effectiveness of the authentication and authorization of the devices. The adapted new embedded layer in the BERT upgrades the performance of the BERT model by understanding the details of the edge device details. The EBERT algorithm represents the various possibilities of the authenticating and authorizing details of the edge devices. The EBERT algorithm comprises of two stage namely embedded stage and encoder stage which is represented in Fig. 15.2.

Embedded Stage. In the embedded phase, every token of the input series was transformed in the embedded vector. Let N be the series length, d be the embedded dimension, the embedded computational complexity is given as $O(N \times d)$ and the four layers of the embedded stage is given as $O(4 \times N \times d)$.

Encoder Stage. The transformer encoder is employed in the EBERT algorithm that comprises of self-attention mechanism with the feed forward neural network. Here L encoder layer, h be the attention head and the self-attention mechanism has $O\left(N^2 \times d\right)$. The computational complexity of the encoder stage is given as $O\left(L \times \left(N^2 \times d + N \times d\right)\right)$. Thus, the significant feature vector details of the edge device data are extracted using EBERT algorithm. The extracted feature vectors are further transferred to the detection stage where authenticating and authorizing in the edge devices are detected based on the extracted feature details.

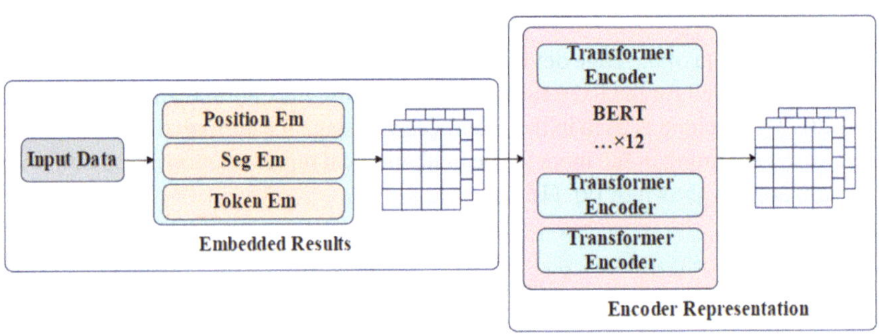

Fig. 15.2 Block diagram of the EBERT algorithm

Authentication and Authorization Detection

The extracted features from the EBERT algorithm are the input for detection stage where the authenticating and authorizing in the edge devices are detected. The RRF (Zhang et al. 2024) is the classifier used for the detection of authenticating and authorizing in the edge devices. The developed RRF model increased its generalization ability toward the unknown data. The model increased its learning ability and recognizes the pattern of the features for detecting the authenticating and authorizing in the edge devices. The RRF incorporated two regularization mechanisms for the control of average depth of the model that increased the authentication and authorization detection ability in the network. Based on the details of the extracted features that RRF model detects where the access in the network is authorized or not that helps to reduce the vulnerable activities in the network at the early stage. After completion of edge device authentication and authorization, the proposed EBERT-RRF model is evaluated using existing metrics.

Experimental Results

The functioning of the proposed EBERT-RRF algorithm is employed in the Python software of version 3.10 and the configuration of system comprises the processor intel core i7, Operating System (OS) Windows 10, Memory 1 Tera Byte (TB), Random Access Memory (RAM) 16 Giga Byte (GB), and Graphic Processing Unit (GPU) 6 GB, respectively. The accuracy, f1-score, recall, and precision are the existing metrics taken for the performance assessment of the developed EBERT-RRF algorithm. The mathematical representation of the performance metrics is given from Eqs. (15.2)–(15.5), respectively,

$$\text{Precision} = \frac{\text{TP}}{\text{TP} + \text{FP}} \times 100 \tag{15.2}$$

$$\text{Accuracy} = \frac{\text{TP} + \text{TN}}{\text{TP} + \text{TN} + \text{FP} + \text{FN}} \times 100 \tag{15.3}$$

$$\text{Recall} = \frac{\text{TP}}{\text{TP} + \text{FN}} \times 100 \tag{15.4}$$

$$F1\text{-Score} = 2 * \left(\frac{\text{Precision} * \text{Recall}}{\text{Precision} + \text{Recall}} \right) \tag{15.5}$$

Here, TP exhibits the true positive numeric, TN shows true-negative numeric, FP gives false-positive numeric, and FN indicates the false-negative numeric. Table 15.1 provides the functioning of the developed EBERT-RRF algorithm. The model has obtained better results with accuracy of 0.925, recall of 0.853, f1-score of 0.859, and

Table 15.1 Performance of the developed EBERT-RRF algorithm

Methodology	Precision	Accuracy	Recall	F1-score
SVM	0.774	0.852	0.752	0.762
NB	0.798	0.876	0.761	0.779
DT	0.816	0.901	0.804	0.809
RF	0.834	0.912	0.817	0.825
Proposed EBERT-RRF	0.867	0.925	0.853	0.859

Table 15.2 Comparative assessment

Methodology	Precision	Accuracy	Recall	F1-score
FL (Loconte et al. 2024)	0.81	0.80	0.80	0.79
XGB (Ehsan et al. 2024)	0.825	0.851	0.824	0.817
Proposed EBERT-RRF	0.867	0.925	0.853	0.859

precision of 0.867, respectively. The support vector machine (SVM), Naïve Bayes (NB), decision tree (DT), and traditional random forest (RF) are the models used for the comparison of developed EBERT-RRF model.

Comparative Assessment

The comparative assessment section provides the improvement of the proposed EBERT-RRF model compared the existing methods used for the detection of authenticating and authorizing in the edge devices in the network. Table 15.2 exhibits the comparative assessment of the developed EBERT-RRF algorithm with the existing methods used for the detection of authenticating and authorizing in the edge devices. The developed EBERT-RRF model has displayed better results with accuracy of 0.925, recall of 0.853, f1-score of 0.859, and precision of 0.867, respectively, compared to the existing FL (Loconte et al. 2024) and XGB (Ehsan et al. 2024) methods.

Discussion

The discussion provides the detailed information of drawbacks of the existing methods for the detection of authenticating and authorizing in the edge device along with the advantages of the proposed EBERT-RRF model. The FL (Loconte et al. 2024) the performance of the authentication system was affected due limited resisting capability of the model for multiple attacks in the network that minimized the security of the IoT data. The XGB (Ehsan et al. 2024) the model interpreted the noises generated during the raining process that led to misclassification of user access details and minimized the authentication performance in the network. The proposed EBERT-RRF algorithm has overcome the existing problems by reducing the generation of noises in the training process and reduced the overfitting problem in the model. The EBERT algorithm extracted the significant details from the data that minimized the dimensionalities of the feature vector and increased the training ability of the model. The developed model has obtained better results of recall of 0.853, accuracy of 0.925, f1-score of 0.859, and precision of 0.867 compared to the existing FL (Loconte et al. 2024) and XGB (Ehsan et al. 2024) methods.

Conclusion

The conclusion says the overview of the detection of authenticating and authoring in the edge devices. The EBERT-RRF is the model proposed for the authenticating and authoring detection of edge devices. The data are collected from smart contrast of various edge devices as the input for detection model. The min–max normalization technique was used in the preprocessing stage to rescale the ranges of the data to normalized value. The EBERT was developed for the extraction of significant details from the data that increased the model training by minimizing the overfitting issue. The RRF classifier was used for the detection of authorization and authentication of edge devices by interpreting the patterns of the extracted feature vector details. The developed EBERT-RRF model has obtained better results of f1-score of 0.859, accuracy of 0.925, recall of 0.853, and precision of 0.867 compared to the existing FL (Loconte et al. 2024) and XGB (Ehsan et al. 2024) methods. Further, the advanced deep learning algorithms can be used for the detecting the authentication and authorization of edge devices that helps to suppress the vulnerable activities in the network and increases the security of IoT data stored in the edge devices.

References

Ahmadi M, Taghavirashidizadeh A, Javaheri D, Masoumian A, Ghoushchi SJ, Pourasad Y (2022) DQRE-SCnet: a novel hybrid approach for selecting users in federated learning with deep-Q-reinforcement learning based on spectral clustering. J King Saud Univ Comput Inf Sci 34(9):7445–7458

Ajao LA, Apeh ST (2023) Secure edge computing vulnerabilities in smart cities sustainability using petri net and genetic algorithm-based reinforcement learning. Intell Syst Appl 18:200216

Almaiah MA, Ali A, Hajjej F, Pasha MF, Alohali MA (2022) A lightweight hybrid deep learning privacy preserving model for FC-based industrial internet of medical things. Sensors 22(6):2112

Alruwaili FF, Asiri MM, Alrayes FS, Aljameel SS, Salama AS, Hilal AM (2023) Red kite optimization algorithm with average ensemble model for intrusion detection for secure IoT. IEEE Access 11:131749–131758

Ehsan T, Sana MU, Ali MU, Montero EC, Alvarado ES, Djuraev S, Ashraf I (2024) Securing smart contracts in fog computing: machine learning-based attack detection for registration and resource access granting. IEEE Access 12:42802–42815

Khashan OA, Khafajah NM (2023) Efficient hybrid centralized and blockchain-based authentication architecture for heterogeneous IoT systems. J King Saud Univ Comput Inf Sci 35(2):726–739

Kumar P, Kumar R, Gupta GP, Tripathi R, Jolfaei A, Islam AN (2023) A blockchain-orchestrated deep learning approach for secure data transmission in IoT-enabled healthcare system. J Parallel Distrib Comput 172:69–83

Loconte D, Ieva S, Pinto A, Loseto G, Scioscia F, Ruta M (2024) Expanding the cloud-to-edge continuum to the IoT in serverless federated learning. Futur Gener Comput Syst 155:447–462

Mazzocca C, Romandini N, Montanari R, Bellavista P (2024) Enabling federated learning at the edge through the iota tangle. Futur Gener Comput Syst 152:17–29

Mishra KN, Bhattacharjee V, Saket S, Mishra SP (2024) Security provisions in smart edge computing devices using blockchain and machine learning algorithms: a novel approach. Clust Comput 27(1):27–52

Razzaq S, Shah B, Iqbal F, Ilyas M, Maqbool F, Rocha A (2023) DeepClassRooms: a deep learning based digital twin framework for on-campus class rooms. Neural Comput Appl 35:8017–8026

Saba T, Rehman A, Haseeb K, Bahaj SA, Lloret J (2023) Trust-based decentralized blockchain system with machine learning using Internet of agriculture things. Comput Electr Eng 108:108674

Sadique KM, Rahmani R, Johannesson P (2023) DIdM-EIoTD: distributed identity management for edge Internet of Things (IoT) devices. Sensors 23(8):4046

Sharadqh AA, Hatamleh HAM, Saloum SS, Alawneh TA (2023) Hybrid chain: blockchain enabled framework for bi-level intrusion detection and graph-based mitigation for security provisioning in edge assisted IoT environment. IEEE Access 11:27433–27449

Tan H (2023) An efficient IoT group association and data sharing mechanism in edge computing paradigm. Cyber Secur Appl 1:100003

Yu B, Tang F, Ergu D, Zeng R, Ma B, Liu F (2024) Efficient classification of malicious URLs: M-BERT—a modified BERT variant for enhanced semantic understanding. IEEE Access 12:13453–13468

Zaidi SAR, Hayajneh AM, Hafeez M, Ahmed QZ (2022) Unlocking edge intelligence through tiny machine learning (TinyML). IEEE Access 10:100867–100877

Zhang X, Shen H, Huang T, Wu Y, Guo B, Liu Z, Luo H, Tang J, Zhou H, Wang L, Xu W (2024) Improved random forest algorithms for increasing the accuracy of forest aboveground biomass estimation using Sentinel-2 imagery. Ecol Ind 159:111752

Chapter 16
Weight Factor-Based Term Frequency–Inverse Document Frequency with Sigmoid Logistic Regression Algorithm for the Authentication of Trust and Privacy Preservation of the Edge Devices

P. S. Abdul Lateef Haroon, N. Dayanand Lal, B. Rajitha, and P. Kiran Kumar Reddy

Abstract The development of the edge devices and network is mainly depending of the trust of the user in the network. The vulnerable activities in the network affected the privacy of the data and minimize the trust of the users. The existing models failed to predict the privacy preserving ability of the edge devices due to limited prediction capability of traditional models. To overcome the existing drawbacks, the weight factor-based term frequency–inverse document frequency with sigmoid logistic regression (WFTFIDF-SLR) model was developed for the authentication of trust and privacy preservation of the edge devices. The data for the analysis were taken collected from the various contrast of the edge devices. The z-score normalization technique was utilized in the preprocessing phased for rescaling the ranges of the data. The weight factor-based term frequency–inverse document frequency (WFTFIDF) for the extraction of important details from the edge device data. The sigmoid logistic

P. S. Abdul Lateef Haroon (✉)
Department of Electronics and Communication Engineering, Ballari Institute of Technology and Management, Ballari, India
e-mail: abdul.lh@bitm.edu.in

N. Dayanand Lal
Department of Computer Science and Engineering, GITAM School of Technology, GITAM University, Visakhapatnam, India
e-mail: dnarayan@gitam.edu

B. Rajitha
Department of Computer Science and Engineering (AI&ML), Vaagdevi Engineering College, Bollikunta, Warangal, India
e-mail: rajitha_b@vecw.edu.in

P. Kiran Kumar Reddy
Department of Computer Science and Engineering (AIML), MLR Institute of Technology, Hyderabad, India

G. P. Ramesh et al. (eds.), *Empowering IoT with Distributed Edge Computing*, IoT and Wireless Communication Networks, https://doi.org/10.1007/978-981-96-9455-6_16

regression (SLR) algorithm for the authentication of trust and privacy preservation of the edge devices. The proposed WFTFIDF-SLR model has shown upgraded results with accuracy of 96.13%, f1-score of 95.19%, recall of 95.89%, and precision of 94.51% compared to the existing privacy protection-based federated deep learning (PP-FDL).

Keywords Authentication of edge devices · Feature extraction · Privacy preservation · Sigmoid logistic regression · Trust · Weight factor-based term frequency–inverse document frequency · Z-score normalization

Introduction

The physical devices that are integrated with the edge computing and located at the boundaries of the network is generally known as edge devices (Hennebelle et al. 2024). The data collected by the Internet of Things (IoT) are collected and processed locally in the edge devices (Awan et al. 2023). The bandwidth of the data transmission is minimized with the help of edge devices that saves the operational cost of the network (Zheng et al. 2023). The framework that is developed for the protection of data and ensures the trust of the users is known as privacy preservation in the edge devices (Wang and Li 2024). The preservation of sensitive information's of the data during processing and transmission process is essential to secure the data and ensures the trust of the users (Jia et al. 2023). The edge devices are running on the batteries and have limited energy resources and memory constraints that creates latency in the data processing and transmission process of the collected data (Alsuqaih et al. 2023). The personal data and sensitive information stored in the edge devices expose to the cyberattacks where the privacy and security of the data is harmed (Abaoud et al. 2023). The trust in the edge device refers to guarantee that the data, their actions and their data are assembled of policies to meet the user's requirement (Odeh et al. 2024). As the technology is rising, the illegal activities in the network like data tampering, unauthorized access, and network traffic are also increasing that affects the security of data and loses the user's trust. The analysis of the edge device data is essential for the detection of privacy preserving ability of the edge devices that helps to ensure the trust of the users (Almagrabi and Bashir 2022).

The literature review exhibits the drawbacks and advantages of the existing methodologies used for the analysis of trust and privacy preservation of the edge devices. Abdel-Basset et al. (2022) presented a privacy protection-based federated deep learning (PP-FDL) mechanism for securing the stored IoT data. The PP-FDL algorithm resisted against the generative adversarial network (GAN) attacks in the network to preserve the privacy of the data stored in the edge devices. The data were categorized based on their weight factors that advanced the generalizing ability of the PP-FDL model and increased the performance of security model to preserve the privacy of data. However, the model stability was affected due irrelevant access in the network that minimized the privacy preserving ability of the network. Shen et al.

(2024) developed a blockchain-based privacy preserving mechanism for IoT network. The Bi-long short-term memory (Bi-LSTM) was the classifier used or the prediction of privacy preserving ability of the network. The deep Q-Learning (DQL) algorithm was used in for the identification of optimal cache decision in the network for trust evaluation that increased the privacy preserving ability of the model. However, the model training was affected with the noises generation that suppressed the privacy preservation prediction ability of the model. Wu et al. (2024) implemented an optimized practical Byzantine fault tolerance (OPBFT) model for preservation of privacy in the network. The two-layered privacy-preserving trust management architecture (PPTMA) mechanism was used for the securing the device data in the network. The blockchain mechanism was incorporated in the network for data storing that advanced the privacy preservation ability of the model in the network. However, the model struggled to handle the multiple attacks in the network due to limited resisting ability that suppressed the performance of privacy preserving mechanism in the network.

Nair et al. (2023) presented a federated learning-based Fed_select mechanism for preserving the privacy and trust of the edge computing network. The Fed_Select algorithm employs substitute minimization mechanism to limit the gradients and applicants in system training for avoiding the vulnerability in the system. The hybrid encryption algorithm was used for increasing the security of the network that enhanced the privacy preserving ability of the model in the edge computing. However, the performance of the model was affected due to dynamic changes in the edge computing network that suppressed the trust and privacy preserving ability of the model. Boopathi et al. (2023) developed privacy preserving data distribution mechanism in the edge computing using optimization algorithm. The firefly grey optimization algorithm was used for the scheduling the tasks in the edge computing. The trust-based multiple encryption technique was used for securing the privacy of the data in the edge computing network that increased the performance of privacy preserving mechanism in the network. However, the latency in the verification process led to unauthorized network access that suppressed the security and privacy of the edge computing data. Hindistan and Yetkin (2023) implemented a hybrid model-based privacy preserving mechanism for the IoT devices. The GAN and differential privacy (DP) models were used for the preserving the sensitive information of the IoT device data. The linear regression (LR) algorithm was used for the prediction of privacy preservation in the network to secure IoT data that increased the performance of security mechanism. However, the performance of the model was affected due to delay in the transmission process where the data packets get lost and minimized data security. The existing methods faced numerous drawbacks during the analysis of trust and privacy preservation of edge devices which are given here as follows: The traditional model struggled to handle the multiple attacks in the network due to limited resisting ability that suppressed the performance of privacy preserving mechanism in the network. The existing model training was affected with the noises generation that suppressed the privacy preservation authentication ability of the model. The traditional models failed to capture the sensitive details of the data that minimized the analysis of trust of the edge devices. The latency in the verification process led to unauthorized network access that suppressed the security and privacy of the edge computing data. The contribution of the proposed WFTFIDF-SLR methodology for the authentication of trust and privacy-preservation of the edge devices is gives below as follows:

- The research aims for the authentication of trust and privacy preservation of the edge devices to secure the stored data against the vulnerable activities in the network where the edge device data are analyzed.
- The z-score normalization was adapted in the preprocessing phased for rescaling the different range value of the data that advance the training performance of the prediction model.
- The weight factor-based term frequency–inverse document frequency (WFTFIDF) algorithm was incorporated in the feature extraction phased to extract the vectorized features of the passcodes that minimizes the overfitting problem occurred in the model training.
- The sigmoid logistic regression (SLR) algorithm was used to analyze the details of the vector features of the edge device data for predicting the trust and privacy preserving ability of the model.
- The weight factor-based term frequency–inverse document frequency with sigmoid logistic regression (WFTFIDF-SLR) model has overcome the existing problem of handling the feature dimensionality and authentication of the trust and privacy preserving capability of the edge devices.

The remaining sections of the research paper are formatted as follows: Section "Proposed Methodology" specifies the proposed WFTFIDF-SLR methodology. Section "Experimental Results" exhibits the experimental results. Section "Discussion" briefs the discussion, and Section "Conclusion" displays the conclusion.

Proposed Methodology

The WFTFIDF-SLR algorithm is the proposed method for the authentication of trust and privacy preservation of the edge devices. The procedure of the prediction model is processed in the four stages such as collection of information of the edge devices, cleaning the collected data in the preprocessing stage using z-score normalization technique, extracting the significant details from the preprocessed edge device passcodes using WFTFIDF method and authenticating the trust and privacy preservation using SLR algorithm. The analysis of edge devices data is essential for monitoring the irrelevant and vulnerable activities in the network that helps to increase the privacy preserving ability of the edge devices that retains the trust of the users in the network. Figure 16.1 exhibits the block representation of the proposed WFTFIDF-SLR methodology.

Fig. 16.1 Block representation of the proposed WFTFIDF-SLR methodology

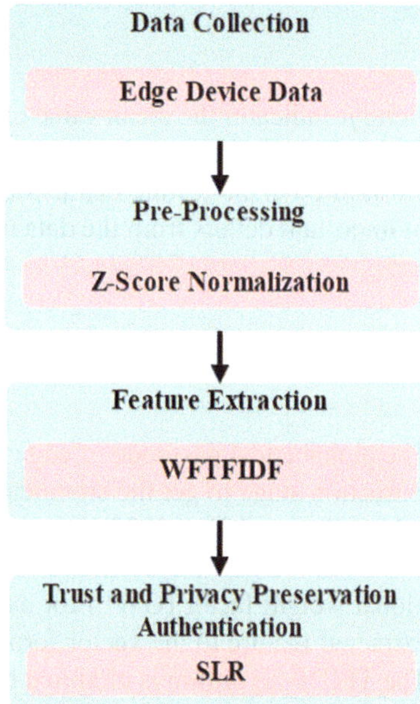

Data Collection

The passcode data are collected from the smart contrast of the various edge devices for the analysis of trust and privacy preservation of the edge devices. The sensor nodes of the edge devices gather the surrounding information. The collected data get processed and aggregated in the edge devices before transmitting to the database. These data are taken as input for the prediction model to analyze the trust and privacy preservation ability of the edge devices. The edge device data are the input for preprocessing stage where the z-score normalization technique is used for rescaling the range values of the data.

Preprocessing

The acquired passcode of edge device data is taken as input for the preprocessing stage to get normalized to increase the training of the model. The z-score normalization (Tihagam and Bhatnagar 2023) technique is used in the preprocessing stage to rescale the range value of the data. The standardization of data leads to improvement in the learning ability of the model to get appropriate prediction of trust and privacy preservation in the edge devices. The mathematical representation of the z-score normalization is given in Eq. (16.1)

$$x_{\text{norm}} = \frac{x_i - \mu}{\sigma} \tag{16.1}$$

Here, μ indicates the mean value of the data, σ exhibits the standard deviation value of the data, and x_i shows the initial data value. The preprocessed data are further transmitted to the feature extraction stage where the model is used for the extraction of important details from the data using WFTFIDF algorithm.

Feature Extraction

The alphabetical passcode of edge device data is considered as input for the feature extraction stage to get the significant information of the data. The WFTFIDF algorithm (Dey and Das 2023) is used for the extraction of edge device data details. The TFIDF algorithm adapted two mechanisms like local weight factor (LWF) and global weight factor (GWF) for extracting the significant details by converting the extracted feature to the vector form that reduces the dimensionality of the features. The TFIDF algorithm is modified based on the weight factors and the mathematical representation of the modified sections id given below from Eqs. (16.2) to (16.6), where the Eq. (16.2) is related to LWF,

$$\text{lwf}_{m,n} = \frac{f_{m,n}}{\sum_p f_{p,n}} \tag{16.2}$$

Here, $\text{lwf}_{m,n}$ provides the details of the LWF for mth term with nth document, $f_{m,n}$ gives the occurred frequency. The Eq. (16.3) is used for modified the GWF.

$$\text{mgwf}_m = \log\left(\frac{\max\{m' \in d\}f'_m}{1 + f_m}\right) \tag{16.3}$$

mgwf_m, provides the details of the maximum GWF for nth document, f_m exhibits the document numbers. The smooth GWF is indicated in the Eq. (16.4),

$$\text{sgwf}_m = \log\left(\frac{N}{1 + f_m}\right) \tag{16.4}$$

Here sgwf_m represents the details of the maximum GWF for nth document and N exhibits the total document value. The updated new GWF value in the model is obtained using Eq. (16.5),

$$\text{mngwf}_m = \text{mgwf}_m + \text{sgwf}_m$$

$$\text{mngwf}_m = \log\left(\max\{m' \in d\}f'_m\right) - 2\log(1 + f_m) \tag{16.5}$$

mngwf_m indicates the modified new GWF nth document. The formula of TFIDF is upgraded based on the modified GWF which is given in Eq. (16.6),

$$\text{wftfidf}_{m,n} = \frac{f_{m,n}}{\sum_p f_{p,n}}\left[\log\left(\max\left\{m' \in d\right\}f'_m\right) - 2\log(1 + f_m)\right] \tag{16.6}$$

Here $\text{wftfidf}_{m,n}$ gives the weight factor based TFIDF value for mth term with nth document. Thus, by varying the global weight facts in the TFIDF algorithm, the significant details of the edge device data are converted to the vector form to increase the training performance of the model. The extracted vector features are further transmitted to the prediction stage where the trust and privacy preservation of the edge devices is analyzed using SLR algorithm.

Trust and Privacy Preservation Authentication

The extracted feature vectors of the edge device data are transmitted to the prediction stage where the trust and privacy prediction features of the edge device data are analyzed. The SLR (Almomani 2023) is the classifier used for the authentication of trust and privacy preservation of the edge device. The LR model is used for the analysis of edge device data based on the details of extracted feature vectors details. To improve the functioning of the LR, a sigmoid function is incorporated in the model for transformation of the LR. The Bernoulli distribution is taken for the distribution of data and the mathematical representation of the Sigmoid LR is given in Eq. (16.7),

$$f_x = \frac{1}{1 + e^{-\text{BO}+B}} \tag{16.7}$$

Here, f_x exhibits the ground truth value. The sigmoid function in the LR is restricted to some values, the mixing of independent base learner helps to reduce the error rate. The mathematical representation id shown in Eq. (16.8)

$$P(h_i(x) \neq f(x) = \varepsilon) \tag{16.8}$$

Here, ε indicates the generalized error rate, $h_i(x)$ gives the base classifier value. Thus, the SLR algorithm predicted the trust and privacy preservation of the edge devices based on the details of the extracted features. The developed WFTFIDF-SLR algorithm has increased the analyzing ability of the prediction model to analyze the trust and privacy preserving ability of the edge devices. After completion of trust and privacy preservation prediction, the WFTFIDF-SLR model is evaluated using existing performance metrics.

Experimental Results

The functioning of the proposed WFTFIDF-SLR algorithm is implemented in the Python software of version 3.11 and the configuration of system includes the processor intel core i7, Operating System (OS) Windows 10, Memory 1 Tera Byte (TB), Random Access Memory (RAM) 16 Giga Byte (GB), and Graphic Processing Unit (GPU) 6 GB, respectively. The proposed WFTFIDF-SLR method performance get assessed using existing metrics which are represented form Eqs. (16.9) to (16.12),

$$\text{Precision} = \frac{\text{TP}}{\text{TP} + \text{FP}} \tag{16.9}$$

$$\text{Accuracy} = \frac{\text{TP} + \text{TN}}{\text{TP} + \text{TN} + \text{FP} + \text{FN}} \tag{16.10}$$

$$\text{Recall} = \frac{\text{TP}}{\text{TP} + \text{FN}} \tag{16.11}$$

$$F1\text{-Score} = 2 * \left(\frac{\text{Precision} * \text{Recall}}{\text{Precision} + \text{Recall}}\right) \tag{16.12}$$

Here, TP exhibits the true-positive numeric, TN displays the true-negative numeric, FP shows the false-positive numeric, and FN represents the false-negative numeric. Table 16.1 exhibits the performance of the proposed WFTFIDF-SLR method. The Naïve Bayes (NB), support vector machine (SVM), K-nearest neighbor (KNN), and traditional logistic regression (LR) are the existing methods used for the comparison. The developed WFTFIDF-SLR model has exhibited greater results with precision 94.51%, f1-score 95.19%, accuracy 96.13%, and recall of 95.89%, respectively.

Table 16.1 Performance of proposed WFTFIDF-SLR method

Methodology	Precision (%)	Accuracy (%)	F1-score (%)	Recall (%)
SVM	90.11	91.52	90.75	91.41
NB	91.56	92.48	92.04	92.53
KNN	92.45	93.76	93.15	93.87
LR	93.10	94.97	93.66	94.24
Proposed WFTFIDF-SLR	94.51	96.13	95.19	95.89

Table 16.2 Comparative Assessment of the proposed WFTFIDF-SLR method with the existing models

Methodology	Precision (%)	Accuracy (%)	F1-score (%)	Recall (%)
PP-FDL (Abdel-Basset et al. 2022)	90.17	93.07	92.09	94.10
Fed-Select (Nair et al. 2023)	NA	94.759	NA	NA
Proposed WFTFIDF-SLR	94.51	96.13	95.19	95.89

Comparative Assessment

The comparative assessment is essential to analyze the improvement of the proposed WFTFIDF-SLR method for privacy preserving of edge devices. The PP-FDL (Abdel-Basset et al. 2022) and Fed-Select (Nair et al. 2023) are the two existing methodologies used for the comparison. The developed WFTFIDF-SLR model has displayed better results having accuracy of 96.13%, f1-score of 95.19%, recall of 95.89%, and precision of 94.51% compared to the existing PP-FDL (Abdel-Basset et al. 2022) and Fed-Select (Nair et al. 2023) methods (Table 16.2).

Discussion

The discussion provides the details of the problems faced by the existing methods during the authentication of trust and privacy preservation of the edge devices. The PP-FDL (Abdel-Basset et al. 2022) and Fed-Select (Nair et al. 2023) methods are used for the comparison of proposed WFTFIDF-SLR model. The PP-FDL (Abdel-Basset et al. 2022) model stability was affected due irrelevant access in the network that minimized the privacy preserving ability of the network. The Fed-Select (Nair et al. 2023) performance of the model was affected due to dynamic changes in the edge computing network that suppressed the trust and privacy preserving ability of the model. The overfitting issue occurred during the training of model due to lack of dimensionality handling ability that suppressed the prediction performance of the model. The existing models failed to authentication of the privacy preserving ability of the edge devices due lack of understanding the sensitive details of the data. The developed WFTFIDF-SLR algorithm has overcome the existing problem by interpreting the details of the feature vector increased the prediction ability of the model. The proposed model also handled the dimensionality issue that minimized the overfitting problem during the model training. The proposed WFTFIDF-SLR model has shown upgraded results with accuracy of 96.13%, f1-score of 95.19%, recall of 95.89%, and precision of 94.51% compared to the existing PP-FDL (Abdel-Basset et al. 2022) and Fed-Select (Nair et al. 2023) methods.

Conclusion

The conclusion briefs the overview of the developed model for the trust and privacy preservation authentication of the edge devices. The WFTFIDF-SRL algorithm is proposed for the authentication of trust and privacy preservation in the network. The passcode data was collected from the smart contrast of the edge devices. The z-score normalization technique was utilized in the preprocessing phased for rescaling the ranges of the data. The weight factor mechanism was adapted in the TFTDF model that formed WFTFIDF for the extraction of important details from the edge device data. The LR incorporated sigmoid function forming SLR algorithm for the prediction of trust and privacy preservation of the edge devices. The developed WFTFIDF-SLR algorithm has exhibited improved results with accuracy of 96.13%, f1-score of 95.19%, recall of 95.89%, and precision of 94.51% compared to the existing PP-FDL (Abdel-Basset et al. 2022) and Fed-Select (Nair et al. 2023) methods. Further, the performance of the trust and privacy preservation authentication model can be improved using optimized features selection techniques that helps for the upgraded analysis of edge device data.

References

Abaoud M, Almuqrin MA, Khan MF (2023) Advancing federated learning through novel mechanism for privacy preservation in healthcare applications. IEEE Access 11:83562–83579

Abdel-Basset M, Hawash H, Moustafa N, Razzak I, Abd Elfattah M (2022) Privacy-preserved learning from non-iid data in fog–assisted IoT: a federated learning approach. Digit Commun Netw 10(2):404–415

Almagrabi AO, Bashir AK (2022) A classification-based privacy-preserving decision–making for secure data sharing in internet of things assisted applications. Digit Commun Netw 8(4):436–445

Almomani A (2023) Darknet traffic analysis, and classification system based on modified stacking ensemble learning algorithms. Inf Syst E-Bus Manag, 1–23

Alsuqaih HN, Hamdan W, Elmessiry H, Abulkasim H (2023) An efficient privacy-preserving control mechanism based on blockchain for E–health applications. Alex Eng J 73:159–172

Awan KA, Din IU, Almogren A, Rodrigues JJ (2023) Privacy-preserving big data security for IoT with federated learning and cryptography. IEEE Access 11:120918–120934

Boopathi M, Gupta S, Zabeeulla AM, Gupta R, Vekriya V, Pandey AK (2023) Optimization algorithms in security and privacy-preserving data disturbance for collaborative edge computing social IoT deep learning architectures. Soft Computing

Dey RK, Das AK (2023) Modified term frequency-inverse document frequency based deep hybrid framework for sentiment analysis. Multimed Tools Appl 82(21):32967–32990

Hennebelle A, Ismail L, Materwala H, Al Kaabi J, Ranjan P, Janardhanan R (2024) Secure and privacy-preserving automated machine learning operations into end-to-end integrated IoT-edge-artificial intelligence–blockchain monitoring system for diabetes mellitus prediction. Comput Struct Biotechnol J 23:212–233

Hindistan YS, Yetkin EF (2023) A hybrid approach with GAN and DP for privacy preservation of IIoT data. IEEE Access 11:5837–5849

Jia D, Yang G, Huang M, Xin J, Wang G, Yuan GY (2023) An efficient privacy-preserving blockchain storage method for internet of things environment. World Wide Web 26(5):2709–2726

Nair AK, Sahoo J, Raj ED (2023) Privacy preserving federated learning framework for IoMT based big data analysis using edge computing. Comput Stand Interfaces 86:103720

Odeh JO, Yang X, Nwakanma CI, Dhelim S (2024) Asynchronous privacy-preservation federated learning method for mobile edge network in industrial internet of things ecosystem. Electronics 13(9):1610

Shen Z, Wang Y, Wang H, Liu P, Liu K, Liu M (2024) Privacy-protecting predictive cache method based on blockchain and machine learning in Internet of vehicles. Veh Commun 47:100771

Tihagam RD, Bhatnagar S (2023) A multi-platform normalization method for meta-analysis of gene expression data. Methods 217:43–48

Wang J, Li J (2024) Blockchain and access control encryption-empowered IoT knowledge sharing for cloud-edge orchestrated personalized privacy-preserving federated learning. Appl Sci 14(5):1743

Wu X, Liu Y, Tian J, Li Y (2024) Privacy-preserving trust management method based on blockchain for cross-domain industrial IoT. Knowl-Based Syst 283:111166

Zheng G, Kong L, Brintrup A (2023) Federated machine learning for privacy preserving, collective supply chain risk prediction. Int J Prod Res 61(23):8115–8132

MIX
Papier aus verantwortungsvollen Quellen
Paper from responsible sources
FSC® C105338

If you have any concerns about our products,
you can contact us on
ProductSafety@springernature.com

In case Publisher is established outside the EU,
the EU authorized representative is:
Springer Nature Customer Service Center GmbH
Europaplatz 3, 69115 Heidelberg, Germany

Printed by Libri Plureos GmbH
in Hamburg, Germany